FAMILY CHEMISTRY: BUILD GENERATIONAL HEALTH

Teach them Babies

Chase Duquesnay EnQi ReaL

Amazon

The characters and events portrayed in this book are fictitious. Any similarity to real persons, living or dead, is coincidental and not intended by the author. None of these books are medical advice, strictly edutainment.

ISBN: 9798334617230

Cover design by: EnQi ReaL

Printed in the United States of America

To the people who invest time and money into other people's children. If not for the Ms. Aprils, Ms. Normas, Ms. Paulettes, Mr. Carters etc...

If not for the Bertha Taylors of the world... this book would not exist! These are just a few of the adults that invested time and money into me as a child.

Dont forget you are someone's baby!!!

LYNETTE JORDAN

CONTENTS

INTRODUCTION

This book is the expanded teacher's edition of sorts! I mean not exactly but...

I know we always need to see what the scientific journals have to say, and I am happy to oblige. We also did such a good job at disguising our teaching of chemistry, some folks was having a Karate Kid moment. I had to highlight what Wax On, Wax Off really is... LOL

It's important, that we provide the Atoms of Self Development for our children. Time is Money and we really dont have either to waste. The best Life insurance...

Remember the saying 'the best offense is a good defense'? Well the best life insurance is the knowledge of Generational Health. You only get 1 custom designer pair of genes in life, learning how to maximize them is your purpose, your real business.

RECEIPTS

We have been all over the place in these books, I do see what Chemistry the books are teaching?

Everything is made from atoms.
Atoms are made of protons, neutrons, and electrons, interacting with Photons.
In Atoms, the number of protons should be the same as the number of electrons.
Atoms (or elements) are organized into the periodic table of elements.
The elements are arranged in rows and columns that have similar qualities.
How atoms combine to form a molecule.
Bonds.
Molecules vs Geometry.
Chemical reactions are based on bonds being created or destroyed, releasing energy or absorbing energy.
Types of chemical reaction:
Decomposition reactions
Displacement reactions
Combination reactions
Exchange reactions
Signs that indicate chemical reactions.
Acids vs Bases/Alkalines
Titration
Homogeneous and Heterogeneous Mixtures
Filtration, Evaporation, Chromatography, Spectroscopy

Carbohydrates, Fats, Proteins
Polymers
DNA

There is a lot of Chemistry, if you caught it all great, we went to great lengths to disguise it... Bwahahahaha

Knowledge is the key, how do you get family to support? They have to believe in the dream! They may also have to believe in getting cut out. Lebron, Floyd Mayweather, Donald Trump, Joe Biden etc... all wealthy people have friends and family that will never want for anything. There are also friends and family that can't even get a phone call or a text. That all depends on wether they help or hurt the cause, if they sneak feed junk food, meat, soda, dairy etc... They are cut out.

Cut out of the family fortune, that is Generational Health. Properly nurtured Generational Health becomes Generational Wealth. God's blessings, walking in your Divine Purpose etc... When we come into the world we have gifts, the things that make us unique are our gifts.
Flip side for Spectrum and Attention Deficit eating just stay pure. Avoid Artificial Colors, Processed sugars and flood your child with Greens, Pulses and Yams...

The BleuMagick Bar should be the source of all 'treats'... Keep in mind we detail exactly how to regulate neurochemistry in Melanin vs Diabetes book two, combined with the Gold book you have a great head start. I think the Generational Health book series offer another advantage. You can bring your child into the knowledge, have your child partner up with you on their nutrition and their specific needs. **Teach your child their strengths and weaknesses, trust them, build with them**.

STANDING ON BUSINESS

Popularized by Drake on the song Daylight, to "stand on business" means to take care of your responsibilities or put your money where your mouth is. You get done what needs to be done and follow through, similar to the term "taking care of business." According to Urban Dictionary, the phrase is defined as: "To take care of your business/obligations; sticking to your values and walking the walk and not just talking the talk.

You make a living with your Genes! We gonna take 'standing on business' from the street. We are going to redefine it.

Standing on Business - bringing out the full potential of your DNA!

Standing - used to specify the length of time that something has lasted or that someone has fulfilled a particular role; erect, upright.

On - (of an electrical appliance or power supply) functioning. Greek 'on' meaning 'being' or the Egyptian city of Light.

Business - the practice of making a **living**.

Let me give you a great example, Jamaica & Kenya. Jamaicans & Kenyans have a high ratio of Fast Twitch Muscle as well as the most active ACTN3 genes. Almost everyone has some fast twitch muscle and over 50% of the population have active ACTN3 genes. Having a high amount of Fast Twitch, with the ACTN3 and a lifestyle to match produces Usain Bolts. I am sure people like Jackie Joyner-Kersee, Floyd and Lebron also have these genetic combinations. On the flip side the XX version or the non-athletic version of this gene, at least athleticism involving fast twitch muscle goes like this: 25% of Asians, 18% of Caucasians, 11% of Ethiopians, 3% of Jamaican and US African Americans, and 1% of Kenyans and Nigerians possessing the XX genotype (Yang et al., 2007; MacArthur et al., 2008; Scott et al., 2010). This means for Kenyans and Nigerians 99% of them have the potential to be elite, 97% of Jamaicans and American Black People etc... Asians have the lowest amount of potential followed by Caucasians...

There is a potential for a counter side regarding slow twitch and endurance sports though. There seems to be a seesaw effect where those who aren't blessed fast twitch wise, are blessed slow twitch wise, balance. The point is you need to pick the right sport for your genes.

You gotta stand on business!!!

Recent genome-wide association studies have successfully identified inherited genome sequence differences that account for 20% of the 50% heritability of intelligence.

You gotta stand on business!!!

Honor your mother and fathers so that your days may be long!!!
You gotta stand on business!!!

This is a two way street, like Floyd Mayweather Jr.. His Father gave him the genes, then taught him how to use those genes!

You gotta stand on business!!!

Why do you think Lebron is making sure both his sons get into the NBA? Why do you think Kobe was making sure his daughter got into the NBA?

People moving in certain circles already have this information.

You gotta stand on business!!!

The ACE I/D gene & the AGT gene codes for angiotensin-1 converting enzyme, part of the renin-angiotensin system responsible for controlling blood pressure by regulating body fluid levels; specifically associated with performance in endurance, but not power.

You gotta stand on business!!!

Variants in two collagen-encoding genes (COL1A1 and COL5A1), a gene involved in connective tissue wound repair (MMP3), and the gene encoding TNC, an extracellular matrix protein, have all been linked to increased risk for tendinopathy.

You gotta stand on business!!!

The PPARA gene has been a good candidate gene to study athletic ability due to its role in lipid metabolism, glucose energy homeostasis and vascular inflammation. It is activated under conditions of energy deprivation, promoting uptake, utilization, and catabolism of fatty acids.

You gotta stand on business!!!

The IGF2 gene provides instructions for making a protein called insulin-like growth factor 2. This protein plays an essential role in growth and development before birth.

You gotta stand on business!!!

The Dystrophin Complex from DMD, the largest known human gene, provides instructions for making a protein called dystrophin. In skeletal and cardiac muscles, dystrophin is part of a group of proteins (a protein complex) that work together to strengthen muscle fibers and protect them from injury as muscles contract and relax. The dystrophin complex acts as an anchor, connecting each muscle cell's structural framework (cytoskeleton) with the lattice of proteins and other molecules outside the cell (extracellular matrix). The dystrophin complex may also play a role in cell signaling by interacting with proteins that send and receive chemical signals. Research suggests that the protein is important for the normal structure and function of synapses, which are specialized connections between nerve cells where cell-to-cell communication occurs.

You gotta stand on business!!!

Many people do just the opposite, coming from poverty they overfeed their children. No one has ever highlighted the fact that **poverty** causes **Calorie Restriction** in children!!! This may contribute to many athletes **Superstardom**.

Creatine concentration (Cr) is greater in **fast twitch muscle fibers**, slow twitch fibers have a greater resynthesis capability due to their increased aerobic capacity. Creatine works with Beta Alanine to facilitate muscle function. Beta Alanine is a naturally occurring amino acid (a non-essential amino acid) not used by the body to make muscle tissue; rather, research has shown that Beta Alanine works by increasing the muscle content of an important compound – Carnosine. In fact, the production of Carnosine is limited by the availability of Beta Alanine. Alanine is an amino acid, while beta alanine is a derivative of alanine. There are many vegetables that are loaded with Alanine! Beta-Alanine can also be formed in the liver from the breakdown of pyrimidine nucleotides into uracil and dihydrouracil and then metabolized into beta-alanine and beta-aminoisobutyrate.

Carnosine is highly concentrated in muscle tissue where its role is primarily to soak up hydrogen ions. Its important to note that fast-twitch muscle fibers have markedly higher carnosine content compared to slow-twitch fibers.

Fast Twitch muscle fibers have more Carbohydrates & Carnosine! Fast-twitch fibers also store a great deal of carbohydrates. For every gram of carbohydrate you store, you also draw about 3 grams of water into the muscle. Thus, bodybuilders who optimize fast-twitch fiber development will obtain a fuller, denser look.

Carnosine (beta-alanyl-L-histidine) is a dipeptide molecule, made up of the amino acids beta-alanine and histidine. Carnosine buffers pH in muscle cells, and acts as a neurotransmitter in the brain. As acidity rises and the muscle pH falls, fatigue sets in. The formation of cross links between proteins responsible for power generation and shortening of the muscle fibers can become compromised. Carnosine buffers that pH decline, and keep muscle working.

Carnosine acts as an **antiglycating agent, reducing the rate of formation of advanced glycation end-products** (substances that can be a factor in the development or worsening of many degenerative diseases, such as diabetes, atherosclerosis, chronic kidney failure, and Alzheimer's disease), and ultimately reducing development of atherosclerotic plaque build-up. Carnosine can **increase the Hayflick limit** in human fibroblasts, as well as appearing to reduce the telomere shortening rate.

A diet high in Phytosterols ie... β-sitosterol. The nomenclature is shown on the right.

- By removing carbon 24, campesterol is obtained.
- By removing carbons 24 and 24, cholesterol is obtained.
- Removing a hydrogen from carbons 22 and 23 yields stigmasterol (stigmasta-5,22-dien-3β-ol).
- By hydrogenating the double bond between carbons 5 and 6, β-sitostanol (Stigmastanol) is obtained.
- By hydrogenating the double bond between carbons 5 and

6 and removing carbon 24, campestanol is obtained.

- Removing carbon 24 and hydrogens from carbons 22 and 23, and inverting the stereochemistry at C-24 yields brassicasterol (ergosta-5,22-dien-3β-ol). Works with Selenium!
- Further removal of hydrogens from carbons 7 and 8 from brassicasterol yields ergosterol (ergosta-5,7,22-trien-3β-ol). Important: Ergosterol is not a plant sterol. Ergosterol is a component of fungal cell membranes, serving the same function in fungi that cholesterol serves in animal cells.

In addition:

- Esterification of the hydroxyl group at carbon 3 with fatty/organic acids or carbohydrates results in plant sterol esters, i.e. oleates, ferulates and (acyl) glycosides.

A diet high in Phytosterols helps to activate your true potential aka helps you **STAN ON BINESS**!

You gotta stand on business!!!

#TEACHINGTHEMBABIES

Mitochondria
Pigments
Lungs
Nose
Eye color
Emotional Baggage
Trauma
Sound perception
Muscle Mass
Hormones
Fat Storage
+Unique Genes
Your unique as a fingerprint physiology.

Typically athletes, trainers and coaches don't calculate these aspects for training programs. The reality is your athletic performance or physical prowess is based on your health. Muscle mass, body fat composition etc... are related to your genetic heritage and that to your ancestors environment.

In the Melanin vs Diabetes movement we understand that Melanocytes are specialized Neurons or that Neurons are specialized Melanocytes. Thinking and Acting are mediated by the two with large Neurons called Nerves connecting the Melanocytes and Neurons to muscle tissue. Thinking and Acting challenges are sports, life and entertainment. Generational Health is Power, more powerful than money and politics. It's the hidden power behind religion, economics and planetary resources.

In the Gold book, the Gold Standard we discuss Nutrimiromics (please read &/or reread that book). The truth is nutrition has to match your genetics blueprint. There is a basic level of nutrition we all share, but optimum health, emotions, thinking and physical performance etc.. all depend on specific complimentary programming. The nervous system, genetic transcription & translation are programs. We all have a unique AlgaRhythm (please read &/or reread that book). Much of our thinking and acting is based on the performance of the nervous system. We drastically overlook Genes, Neurons, Nerves and Melanocytes in Athletics.

Generational Health is largely based on the specifics of your genetic code, the Single Nucleotide Polymorphisms you carry. For every genetic mutation that causes disease or weakness, there is a mutation that creates an abnormal gift! Cognitive, Proprioceptively, Mechanosensitively, Athletically, Emotionally, Empathetically, Extra Lucid Dreaming, Extra Sensory Acoustic Perception etc... Think about the steady hands of artist or surgeons, everyone doesn't have the hands or coordination to be a surgeon, even if they have the mind for it. The Ancient story of Osiris has overtones that you can't escape. **Osiris** is known as the **Perfect Black**, depicted with **Green Skin** and fermented fruit sugar for **Blood**. When you acknowledge that black people have more type 2b fast twitch muscle, it should give you the chills! Type 2b fast twitch muscle can not use oxygen to provide energy only the fermentation process, only fermented fruit sugar. The odds on the perfect black having blood of fermented fruit sugar with zero knowledge of anatomy are a trillion to one. Type 2 diabetes could easily be Type 2 Muscle Disease. In Melanin vs Diabetes book one we renamed Diabetes, Melanin Concentrating

Hormone Disease.

Acanthosis Nigricans: When the body produces too much ineffective insulin, it increases the amount of melanin or pigment in the skin, which is a condition known as acanthosis nigricans.

That alone is crazy when you think about the type 2 diabetes statistics!

Thinking, Feeling and Acting. The speed at which you can process information and create the correct physical, mental or emotional response. In many situations the correct response is some combination of physical, mental and emotional actions.

This all comes down from the Heart & Brain to the Aorta/Was Sceptre and the Spine/Djed Pillar. Pigment and Fiber are the keys to speeding up information processing.

The truth is that we have outlined the foods and supplements for up-regulating nerve transmissions, impulses, bioelectrical muscle tissue, neural communication, electromagnetic and luminous tissue already in the previous books:
Food Basics
HydroChemistry
Lymphatic Immunity
Food Chemistry (the Constitution)
Kinesiology Chemistry (Declaration of Independence)
Kitchen Chemistry
Orthorexia
Autodidact

This book and subsequent series is about making you and you family aware of your genetic wealth. Fast twitch gifts, slow twitch gifts, cognitive gifts etc... We want to inspire you to find these gifts in your children and foster them!

All men are not created equal! You need to learn who you are and who you make babies with, so you will know your babies!!! Once you know their strengths and weaknesses they can be applied a million ways for a successful life. This is the true meaning g of walking in divine purpose.

There are many things that people may be looking for that we have addressed in a different way than what they are accustomed ie.. motor engrams. We have discussed Martial Arts throughout all of our discussions. If you know what motor enigmas are, and the practice of martial arts, you could easily call martial arts motor engram encoding science. I would definitely say there is a Omega 3 problem though! Especially since Melanocytes & Neurons haven't been properly recognized as twin cells and 3,000 new neurons were just discovered... That means we can't fully understand the myelination, ATP vs GTP etc... I have said for decades that there is something creepy about the sale of dirty, industrial chemical laden, fried grease in the urban communities! Fish places, burger joints, Chinese spots, chicken spots etc are really only selling fried rancid grease!!! The people buying fried chicken & fried fish don't like raw meat. They are in fact hooked on that grease and they don't understand the origin of those cravings.

Motor Engrams - Memorized motor patterns used to perform a movement or skill, that are stored in the motor area of the brain. Motor skill acquisition occurs through modification and organization of muscle synergies into effective movement sequences.

The conversion ratios of substrates into nutrition is primary to sports nutrition and cognitive fitness. Seeing the body as a integrated system of electronic components opens a doorway to better physical and cognitive performance as well.

Thought, Feeling, & Actions must take place in lightning speeds. Neuromuscular coordination is not just 1 muscle, its hundreds of skeletal muscles, thousands of smooth muscle, with hundreds of thousands into the millions of heart muscles, coordinating thoughts, feelings and actions.

There has to be maat between the blind rage of the sympathetic, and the cool calculated energy of the parasympathetic portions of the autonomic nervous system. You don't want any stems (stimulants) as apart of your nutrition. The aromatic amino acids and glutamate are required in excess for scholars &/or athletes. Many people swear by coffee, pre-workouts and energy drinks but I don't. The body makes everything it needs, anything you take will slow down internal production.

Dopamine (tyrosine) is the most important in our Research, as we have previously detailed how it regulates all brain function. Please read &/or reread EnQi and the Brainwave as well as the You got some Nerve books. This may sound crazy but reading and having thought provoking conversation is a dope way to boost dopamine. Dopamine comes from the Neurons that are melanated! Love and Fear are the only two pathways to get 'in the zone'. Fear works to slow down time just as Dopamine does but, the fear pathways deplete sugar supplies. Doing what you Love slows time without the stress hormones. Serotonin is the time flies neurotransmitter, it doesn't allow focus. In many case fatigue is associated with Low Dopamine and High Serotonin levels.

Dopaminergic, MCH (we have spent whole books on Melanin Concentrating Hormone) and Cholinergic Neurons are the big for muscle/cognitive performance.

Choline is gas for muscles, choline + acetyl-CoA = Acetylcholine. Acetylcholine is the neurotransmitter used at the neuromuscular junction—in other words, it is the chemical that motor neurons of the nervous system release in order to activate muscles. Acetylcholine is synthesized in certain neurons by the enzyme choline acetyltransferase from the compounds choline and acetyl-CoA. Cholinergic neurons are capable of producing ACh. Acetylcholine functions in both the central nervous system (CNS) and the peripheral nervous system (PNS). In the CNS, cholinergic projections from the basal forebrain to the cerebral cortex and hippocampus support the cognitive functions of those target areas. In the PNS, acetylcholine activates muscles and is a major neurotransmitter in the autonomic nervous system.

Remember that Acetyl-CoA is produced from Fat & Carbohydrates + Vitamin B5.

Big mistake many people around the world in every aspect of life make, SUNLIGHT. The cutting edge science is just getting to Vitamin D. The problem is the world hasn't accepted our diet as the premier diet around the world. In the Photochemistry book (please read &/or reread that book) we outline how all light penetrates the tissues to different depths. Vitamin D is just the beginning, light changes the behavior of mitochondria, various hormones, all of the pigments in your body, the blood vessels are lined with Melanopsin and carry food pigments...

In a world where electricity, light and magnetism are not included in functional anatomy, pigment is discarded. Thanks to Robert O Becker and Dr. Sebi the concept of the body being electric is beginning to seep in. At the height of this revelation, I became the black sheep by going against the grain with a personal revelation. The body is not electric, it only uses electricity to facilitate it's luminosity. Electricity is the source of energy for Light & Magnetism. Our mantra became 'Plasma carries Electricity, moving Electric Fields create Magnetism, alternating Electric & Magnetic Fields create Light'. This understanding of the body is how I healed Amber's broken neck and spine etc... Flooding the body with plant pigments creates magic, flooding the body with fiber creates magic! Flooding the body with Light via Pigment and Water via Fiber, creates magic! This magic translates to Transcription, Translation etc... the Declaration of Independence book is a treasure trove of jewels regarding this magic.

Vitamin D is a group of active nutrients, every pigment we consume becomes a group of active nutrients. Endogenous pigments function with exogenous pigments to facilitate our particular degree of thinking, feeling and acting.

What does the endocrine system mean in the world of holistic science? Chakras!

Seeing as how white folks gravitate more to Naturopathic Science, and black folks gravitate more to Holistic Science, it's no wonder to me why black people are the leaders of all disease metrics!

Naturopathic Science is Holistic Science, Holistic Science is Naturopathic Science, in theory. The problem is that Holistic Science carries a bunch of gobbledegook that is nonsensical! We have spent millions of hours, millions of dollars studying nonsense. Chakras don't absorb, store, transduce, redistribute, convert light, pigment does. The endocrine system produce a wide variety of what I call mobile pigments. The reason is most of our hormones are built from/with the very same aromatic structures as pigment.

Thinking (Cognitive)) - Feeling (Emotional) - Acting (Physical)

Neurons - Melanocytes - Nerves

When we fully understand these relationships we will understand Anabolic and Catabolic better. You have 1 body, all the systems function together. Higher testosterone levels work for the Cognitive & Emotional faculties just like your Physical faculties! The type 2b fast twitch muscle that is a curse for black people thus far, may be our biggest blessing. We have been creating millionaires and billionaires through this Generational Health. We just haven't kept the spotlight on our Imhoteps, Lewis Latimers, George Washington Carvers, Henry T Sampsons etc...

PQQ, **Lion's Mane**, **Nicotinamide Riboside**, **TMG**, **CoQ10**,

Urolithin A, **Creatine** are all good for athletics but we are missing the obvious. Billions are generated yearly for physical performance but we overlook the Heart & Brain. There are thousands of times more mitochondria in the Heart & Brain than in the body. This highlights burning fat and sugar in tandem, YouTube gurus and even textbooks treat the subject as if one happens without the other. Everything is happening in the body at once, EVERYTHING. Fats & Carbs metabolized, proteins are being built, cells are destroyed and recycled etc... There is a daily turn over of about 50 billion cells!!! The Kitchen Chemistry, Orthorexia, Autodidact changed the way we see plants as food, we need 7 - 15 servings a day, much more than the RDA!

#GreenManChallenge

#10kaDay

#BleuMagickHeartChallenge

We have to build reserves of nutrition in our body, no matter what biochemical pathways the body needs for the thinking, feeling, acting matrix. If we are loaded, reserve wise, nutrient reserve plays just as crucial a role as VO2 Max or Hydration in physicality, emotional intelligence and cognitive performance. We have to learn, the brain loves sugar and the type 2b muscle tissues love sugar, the heart loves fats... Can you think or exercise without the Heart?

Metabolic Flexibility is a requirement for high performance. Different sports require different sets of nutrition, Marathon Runners vs Sprinters, a chess player vs an IT specialist, are great examples. I am hoping that you are learning as you are teaching!

Neurogenesis, Melanogenesis & Myogenesis are all tied to one another! Holistic doesn't mean chakras and other goofy ideas of pseudo spirituality, holistic means treating the body as a whole system of systems. Training isn't just GAINS or Mass, there is a point where mass contributes to longevity, contributes to sport, contributes to endurance and there's a cut off where it hurts function, hurts longevity.

What is the mass trained for? Have you created the right AlgaRhythms within the neuromuscular pathways? What percentage of the Mass we are gaining is muscle vs fat? What percentage of the fat is Brain, Myelin, Wat vs Bat etc…?

PRE-BLACK LINES MATTER DICTIONARY

Agar

Agar is primarily composed of carbohydrates (mostly fiber) and has very low levels of vitamins. It is not a significant source of essential vitamins such as vitamin A, vitamin C, or the B vitamins.

Agave

Agave syrup, which is extracted from the agave plant, primarily contains carbohydrates in the form of fructose and glucose. It is not a significant source of essential vitamins such as vitamin A, vitamin C, or the B vitamins. Despite its lower glycemic index, excessive consumption of agave syrup may contribute to weight gain and metabolic issues due to its high fructose content. Fructose metabolism differs from glucose and can lead to adverse health effects when consumed in large amounts.

Amaranth
Amaranth is rich in vitamins A, C, E, and B vitamins like folate, riboflavin, niacin, thiamine, and vitamin B6. Amaranth is highly nutritious, offering protein, fiber, minerals, and antioxidants. Amaranth's antioxidants protect cells from damage, potentially lowering the risk of chronic diseases.

Antiparticle

A subatomic particle with the same mass as a particle of normal matter but with opposite properties. Antimatter is made up of antiparticles.

Allyl

Derivatives inhibited carcinogenesis in the stomach, esophagus, colon, mammary gland, and lungs of experimental animals and improved immune function, reduced blood glucose, and conferred radioprotection and protection against microbial infection. Allium vegetables have been used for medicinal purposes throughout recorded history to increase longevity, stamina, and strength, and as an antiparasitic agent, antiseptic, antimicrobic, antipyretic, and analgesic.

Allspice

Pimenta dioica contains numerous phytochemicals including phenolics, vanillin, eugenol, and terpenoids. In traditional medicine, allspice has been used to treat hypertension, inflammation, pain, diarrhea, fever, cold, pneumonia, and bacterial infection.

Almond

Almonds are a source of protein, alpha-tocopherol, manganese, magnesium, copper, phosphorus, and riboflavin, in addition to fiber, phytosterols, and polyphenolic compounds, such as proanthocyanidins and lignans. Almond fat is mostly monounsaturated; approximately 50% of an almond's weight is fat. Per 8 oz, almond milk supplies 450 mg of calcium and 3.75 μg (150 IU) of vitamin D compared to cow's milk that supplies 276 mg of calcium and 3 μg (124 IU) of vitamin D. When consumed as part of a low saturated fat diet, low cholesterol diet, 2.5–3.5 oz (70–100 g) of almonds reduced total cholesterol by 4–11% and LDL by 7–12%, according to the results of five small human studies, two of which also found a 1.7% to 3.5% increase in HDL cholesterol.

Anise

Anise seeds are used as a spice and to flavor liqueurs. Though anise seeds are commonly consumed after meals to freshen breath and to induce burping, which is thought to occur because a constituent in anise reduces the surface tension of the stomach contents, resulting in gas bubble coalescence and release. Anise demonstrated antimicrobial, antifungal, antiviral, antioxidant, immunostimulant, muscle relaxant, analgesic, and anticonvulsant activity, reduced morphine dependence, antagonized Helicobacter pylori, and exerted cytotoxic activity in human prostate cancer cells in vitro.

Anthocyanins

Purple or red plant antioxidant food pigments, found in foods such as berries, black currant, blueberry, cherry, cranberry, eggplant, lingonberry, mulberry, lettuce, and strawberry. Used in folk medicine to treat liver dysfunction, hypertension, vision disorders, microbial infections, diarrhea, and other disorders.

Antioxidant

Antioxidants may interfere with oxidation by inactivating a prooxidant compound, scavenging free radicals, acting as chelators to inactivate metal catalysts, repairing oxidative damage, or stimulating the activity of antioxidant enzymes. All plant foods are sources of antioxidants, which include carotenoids; vitamin C, vitamin E, and selenium; isothiocyanates; and phenolic compounds (flavonoids, stilbenes, phenolic acids, and lignans).

Apple

Member of Rosaceae family and popular snack fruit that is available dehydrated, dried, and as sauce, juice, and cider. There are thousands of varieties of apples having fairly similar nutritional profiles. Apples and apple products have been associated with beneficial effects on the risk, markers, and etiology of cancer, cardiovascular disease, asthma, and Alzheimer's Disease.

Apricot

Small stone-fruit member of the Rosaceae family that supplies, per four apricots (a standard serving): 2.8 g of fiber (10% DV), 14 mg of vitamin C (15% DV), and 2696 IU of vitamin A (89% DV) and phytochemicals includ- ing flavonoids. In folk medicine, apricots have been used to treat hem- orrhage, infertility, eye inflammation, and spasm.

Artichoke

Artichoke hearts contain both soluble and insoluble fiber and are a good source of fiber, vitamin C, folate, and magnesium. Artichoke's phytochemicals include caffeic acid and flavonoids.

Arugula

A good source of vitamin K77 and glucosinolates. Eruca sativa seed extract has been used to treat skin disorders in traditional Middle Eastern medicine. Arugula seeds are used to grow microgreens used in salads and as a garnish.

Asian Pear

A source of malaxinic acid, a phenolic acid compound found mainly in pears, and they contain 4 g fiber (14% DV) per medium fruit. In laboratory research, Pyrus pyrifolia cv. Shingo (Korean pear), a specific variety or cultivar of Asian pear, stimulated two key alcohol-metabolizing enzymes involved in alcohol detoxification.

Asparagus

Asparagus is a good source of vitamin K and folate, and contains numerous phytochemicals, including flavonoids and saponins. The mineral content (copper, iron, zinc, manganese, cal- cium, magnesium, sodium, potassium, and phosphorous) is higher in green asparagus than in other varieties, and nutrients are generally richest in the tips of the spears. In an 8 week animal study, an asparagus extract reduced total and LDL cholesterol and raised HDL cholesterol in mice fed a highfat diet to induce hyperlipidemia; in addition, antioxidative effects and normalization of animals 'liver function tests were attributed to the asparagus extract.

Avocado

Avocado is an excellent source of folate, is high in monounsaturated fat, and is a source of the phytochemicals lutein, xea- xanthin, and beta-sitosterol, supplying 114 mg of beta-sitosterol per cup of commercial variety avocado and 175 mg per cup of California avocado. The oils in avocado may theoretically promote absorption of fat-soluble vitamins that are consumed in the same meal.

Acid-base reaction

A particular type of chemical reaction in which an acid neutralizes a base.

Adenine

One of the four nucleic acid bases that make DNA.

Alkali
The group of elements in the far left-hand column of the period table.

Alkaline
The group of elements in the column just to the right of the alkali metals.

Amine
Two hydrogen atoms connected to a nitrogen atom which is connected to another molecule.

Amino acid
A molecule with a central carbon atom that has a carboxylic acid group, an anime group, and an "R" group attached.

Amylase
The protein that breaks down amylose into individual glucose molecules.

Amylose
A common unbranched polysaccharide found in plants.

Amylopectin
A common branched polysaccharide found in plants.

Atomic number
The number located in the upper left-hand corner near the atomic symbol. It tells the number of protons in an atom's nucleus.

Atomic weight
The combined weight of an atom's protons and neutrons.

Atoms
A fundamental unit of all matter composed of protons,

neutrons, and electrons.

Axis
The main vertical or horizontal line on a plot.

AIDS
A disease caused by the HIV virus. It attacks the body's immune system. AIDS stands for "acquired immune deficiency syndrome".

Algae
Eukaryote autotrophs that take energy from the Sun.
Algae range from single celled microorganisms to multi celled seaweeds.

Amoeba
A tiny, single celled eukaryote that can change its shape. Amoebas are mostly protists but can also be fungi, algae, and animals.

Antibiotics
Drugs designed to kill microbes or stop them from growing.

Antibody
A special protein produced by the body that latches on to bacteria and viruses, "flagging" them as invaders for the immune system to deal with.

Accumulator
A device such as a rechargeable battery that stores energy.

Additive
A way of manufacturing products by building up layers, adding one on top of the other the opposite of manufacturing objects by removing material.

Alloy

A metal blended with other metals of nonmetals to make a material with desirable properties. Alloys include bronze, solder, and steel.

Analog

When an electric signal has the same shape as the voice or picture being transmitted.

Atom

A tiny particle of matter, made of protons, neutrons, and electrons; the smallest component that can take part in a chemical reaction.

Banana

Popular snack fruit frequently recommended to maintain or restore blood potassium levels because it is a good source of potassium (487 mg of potassium, 10% DV) per large, 8–9 inch fruit; also a good source of fiber (3.5 g, 12.5% DV) and vitamin C (11 mg, 12% DV), and an excellent source of vitamin B6 (0.449 mg, 29% DV). Bananas contain serotonin, a precursor of melatonin. A low glycemic index food.

Barley

Whole grain and a good source of fiber, barley is used to make vegetable barley soup (3.9 g fiber [14% DV] per 1 cup). Contains β-glucan, a soluble fiber, in addition to B vitamins (thiamin, riboflavin, niacin, vitamin B6, folate, pantothenic acid, and biotin), iron, zinc, phosphorus, and magnesium.

Basil, sweet

Mint family herb that contains phenolics, terpenoids, and steroids including stigmasterol and β-sitosterol, it supplies significant amounts of vitamins A 633 IU (20% DV) and K 50 μg

(41% DV).

Baryon

A subatomic particle made of three quarks. Protons and neutrons are baryons.

Black hole

The densest thing in the universe. So massive that not even light can escape its gravity.

Bay Leaf

Bay leaves contain small amounts of vitamin A, vitamin C, and vitamin B6.Bay leaves contain compounds that act as antioxidants, protecting cells from damage. Research suggests bay leaves may help improve insulin sensitivity and regulate blood sugar levels. Bay leaves, besides adding flavor to dishes, contain vitamins and beneficial compounds that may support antioxidant activity, digestive health, and blood sugar regulation.

Beet

Root vegetable that supplies iron (<3% DV) and is a good source of folate (18% DV) per ½ cup. Beets have high levels of polyphenols and betalaines that exert antioxidant effects. Beet greens are good sources of calcium (12% DV), iron (15% DV), and vitamin E (17% DV), and excellent sources of magnesium (23% DV) and vitamin K (580% DV) per one cup.

Beta-carotene

Vitamin A precursor present in rich amounts in yellow- and orange- fleshed fruits and vegetables, certain red fruits and vegetables, and dark green vegetables. Good sources include carrots, pumpkin, and spinach. Cooking a fruit or vegetable may enhance its β-carotene bioavailability.

Beta-glucan

Bioactive polysaccharide found naturally in foods that contain yeast, fungi (e.g., mushrooms), certain bacteria, seaweeds, and grains, such as oats and barley. The regular consumption of β-glucans contributes to the maintenance of normal blood cholesterol concentrations. Yeast-derived β-glucans dosed at 7.5 g twice daily, significantly reduced total cholesterol concentrations by 6–8% in patients with hypercholesterolemia after 7–8 weeks of treatment.

Beluga Lentil

Beluga lentils are rich in vitamins B1, B5, B6, B9 (folate), and K.They help lower cholesterol and regulate blood pressure. Beluga lentils have a low glycemic index, aiding in blood sugar regulation and insulin sensitivity.

Biotin

Water soluble vitamin that functions as a coenzyme in energy metabolism that is widely distributed naturally in foods such as meats, cereals, and fruits. Biotin is also synthesized in the intestine though not in a form that contributes significantly to absorbed biotin perhaps less than 20% of a person's daily requirement for biotin is from enterically synthesized biotin.

Bitter Orange

Bitter orange has been used in traditional or folk medicine for weight loss; to relieve nausea, indigestion, constipation, heartburn, loss of appetite, and nasal congestion; topically, it is used to treat fungal infections, such as ringworm and athlete's foot.

Black Currant

High-vitamin C berry that is a significant source of anthocyanins. The effect of black currant anthocyanins on the progression of glaucoma was evaluated in a placebo-controlled, double-blind trial n = 38 glaucoma subjects) in which subjects taking eye drops were randomized to receive either oral black currant anthocyanins (n = 19) or a placebo (n = 19) for two years.

Black Pepper

Peppercorn of the Pipernigrum vine that is among the most-used culinary seasonings and considered “the king of spices” throughout the world due to its pungent principle piperine. Piperine exerted beneficial, protective effects against inflammation and alveolar bone loss, supported bone microstructures and prevented collagen fiber degradation in an experimental periodontitis study. Experimental studies have also found Pipernigrum and its secondary metabolites to have antiapoptotic, antibacterial, anti-inflammatory, and hepatoprotective properties.

Blackberry

High-fiber (3.8 g, 24% DV), high-vitamin C (15 mg, 16% DV) berry (values per 1/2 cup) that has an edible but bitter solid center. It is not the same as a black raspberry which is hollow. Contains polyphenolic compounds such as ellagic acid.

Black rice

Black rice is rich in vitamins E, B1 (thiamine), B2 (riboflavin), B3 (niacin), and B6.Black rice contains antioxidants that protect cells from damage. Compounds in black rice have anti-inflammatory properties.It may help lower cholesterol and blood pressure, promoting heart health.

Blueberry

Blueberries supply vitamins A, C, and B6, thiamin, riboflavin, folate, potassium, and fiber, in addition to being abundant in flavonoids. Blueberries 'high antioxidant capacity relative to other fruits have made them the focus of neurological research examining blueberry flavonoids in protecting the brain from oxidative stress.

Bok Choy

Bok choy is a good source of calcium, vitamin K, potassium, and vitamin C, in addition to glucosinolates such as isothiocyanates and indoles.

Boron

Essential trace mineral highest in avocado, peanut butter, peanuts, prune, grape juice, chocolate powder, wine, pecans, granola raisin cereal, and raisin bran cereals.

Boysenberry

Boysenberries are rich in polyphenolic compounds (flavonoids), the main antioxidant group in berries. vascular health parameters found positive changes in flow-mediated dilation of the brachial artery and changes in systolic blood pressure consistent with reducing cardiovascular risk.

Brassica Vegetables

Genus consisting of many common species of cabbage family vegetables that exclusively contain glucosinolates, such as sulforaphane and indole-3-carbinol, and glucosinolate breakdown products called isothiocyanates (ITCs), in addition to polyphenols, terpenes, anthocyanins, coumarins, antioxidant vitamins C and E and carotenoids, and beneficial antioxidant enzymes. Brassica is the largest and most widely consumed group of plants worldwide.

Brazil Nut

Brazil nuts are a particularly rich source of selenium, supplying 95 mcg of selenium (172% DV) per 1 nut. High in fat, each 5 g nut provides 3 g of monounsat- urated and polyunsaturated fat and 30 calories, plus nutrients including thiamin, vitamin E, magnesium, copper, and manganese. Brazil nuts have been used to correct selenium deficiency.

Broccoli

A good source of fiber, magnesium, potassium, vitamin C, folate, and carotenoids, it is also a good source of calcium: a broccoli stalk with floret (150 g) supplies 70 mg (5% DV) of calcium. A laboratory study found that broccoli soluble fiber prevented E. coli translocation in Crohn's disease cells which may have implications in preventing Crohn's disease by limiting harmful bacteria's access to vulnerable intestinal cells.

Brussel Sprouts

A source of vitamin K, lutein, xeaxanthin, and the antioxidant glutathione.Brussels sprout diet during tumor initiation exhibited a 67% incidence of fibroadenomas, whereas, rats fed the casein-cornstarch diet during initiation, but switched later to the Brussels sprouts diet, showed over a 90% incidence of adenocarcinomas.

Top Foods that are excellent for boosting bone health:

1. Leafy Green Vegetables : Kale, spinach, collard greens, and broccoli provide calcium, vitamin K, and other nutrients important for bone health.

2. Salmon and other Fatty Fish : Rich in vitamin D and omega-3 fatty acids, which help enhance calcium absorption and bone strength.

3. Almonds : High in calcium and also provide magnesium, which supports bone mineral density.

4. Tofu : Made from soybeans, tofu can be a good source of calcium if prepared with calcium sulfate.

5. Oranges : High in vitamin C, which is important for collagen production and bone mineralization.

6. Seeds : Chia seeds, flaxseeds, and sesame seeds provide

calcium, magnesium, and other minerals essential for bone health.

Incorporating these foods into your diet can help support and maintain strong bones over time.

The Top Regular Foods That Boost Your Brain Health

Blueberries

In adults, the hippocampus (the part of the brain that's a center for memories, emotions and spatial orientation) loses about 1% of its volume each year. Research has shown that eating blueberries, which contain brain-preserving phytochemicals, can prevent and possibly even reverse the shrinkage that's associated with the onset of dementia and Alzheimer's disease.

Green leafy vegetables

Green leafy vegetables, like spinach, kale, escarole, collards, and arugula, are the most important of all veggies to protect the brain from cognitive decline. A study showed that eating leafy greens delayed cognitive aging by 11 years. Try to consume one cup raw or a half cup cooked greens each day. Many of the brain-boosting phytonutrients in leafy greens are fat soluble and are best absorbed when served with healthy fats like olive oil and avocados.

Beans

Beans are high in folate and B-vitamins, and research suggests these nutrients help prevent or slow brain shrinkage. Eat any bean you like—they're all great for you. Chick peas, northern white beans, and black beans are great in salads and provide a healthy base for dips like hummus. Pureed cannellini beans also make a great base for a sauce. What's more, the high soluble fiber in beans feeds the good bacteria in your gut that in turn lowers inflammation throughout your body.

Dark chocolate

As you eat dark chocolate, the helpful nutrients (flavonoids) immediately begin to improve blood flow to your brain. These nutrients can boost your working memory and problem-solving skills. They also increase the amount of nitric oxide that is produced by the cells that line the inside of all blood vessels. This nitric oxide has an anti-inflammatory effect. It's best to cook with unsweetened dark chocolate powder because it's packed with helpful flavonoids and has no sugar.

Coffee and tea

Both coffee and tea can decrease the calcification of blood vessels. Tea is even better, since it has less caffeine than coffee. Tea also contains a phytochemical called theanine, which has both a calming and stimulating effect, and that also increases dopamine, a feel-good brain chemical.

Turmeric

Bright yellow turmeric is a spice commonly found in Indian curries and mustard, but you can easily add it to any number of dishes. You can sprinkle it into anything you cook, as the flavor is mild, and it blends well with almost all soups and sauces.

Turmeric helps remove a specific plaque in the brain that contributes to Alzheimer's disease. But, eat the real spice and avoid the supplements. Only the spice itself has the antioxidant properties that are most effective in boosting brain health.

Tomatoes

Cooked tomatoes contain high amounts of lycopene, an antioxidant found in many red, orange, and yellow fruits and vegetables. Lycopene reduces oxidative stress, which is damaging to your brain. Tomato sauce and sofrito, two popular Italian and Latin-American foods, are great ways to load up on lycopene.

Pistachios

Pistachios are very high in vitamin E, which has well-documented brain-protective qualities. The natural oil in pistachios also can prevent brain inflammation, and some studies even suggests it can reduce frontal lobe shrinkage in those who've experience certain brain injuries. Keep your portion size to a small handful.

Bond
The attachment between two atoms in a molecule. A chemical bond can be either covalent or ionic.

Bioluminescent
Describes living things that can produce light. A number of marine animals and insects can do this.

Cabbage

Both white cabbage, which is pale green, and red cabbage, which is magenta and white, per 1 cup, are good sources of vitamin C

(32 mg, 11% DV) and vitamin K (67 µg; 55% DV), and supply glucosinolates and antioxidants; additionally red cabbage is a good source of vitamin A (993 IU, 33% DV).

Caffeine

Caffeine improves mental alertness, prevents fatigue, and may relieve a simple headache, though caffeine withdrawal can cause headache and fatigue. Caffeine is an ergogenic aid that increases work output; it increases the catabolism of serum free fatty acids and muscle triglycerides to enhance performance.

Calcium

Major mineral found primarily in the skeleton but necessary in the blood and soft tissue for nerve transmission, muscle contraction, blood clotting, blood pressure maintenance, and other physiological processes. Good sources, in addition to dairy products, include many dark green vegetables, such as bok choi and broccoli.

Calamari

Calamari provides essential vitamins, particularly B12, B6, B3, B2, and C, along with beneficial omega-3 fatty acids and minerals. Scientifically, it supports heart health, brain function, and overall wellbeing, making it a nutritious choice when included in a balanced diet.

Cantaloupe

Orange-fleshed melon that contains vitamin E (0.9 mg, < 1% DV) and fiber (1.6 g, 5% DV), and is a good source of vitamin C (65 mg, 72% DV) and an excellent source of vitamin A (5986 IU, 199% DV) per 1 cup. Cantaloupe also supplies carotenoids, phenolics, and terpenoids.

Canthaxanthin

A type of xanthophyll carotenoid chemically related to the vitamin A pre- cursor betacarotene and presumed to be an antioxidant. Found naturally in plant and animal foods such as mushrooms and trout and added to foods such as farmed salmon, drugs, and cosmetics as a colorant.

Caper

Capers contain numerous phytochemicals including phenolics, tocopherols, sterols, alkaloids, glucosinolates, and fatty acids. Traditionally used as a diuretic, astringent, and antidiabetic, antihyperlipidemic, and antirheumatic agent.

Carambola

Also called star fruit. Yellow, star shaped fruit of the Oxalidaceae family that is popular in Asian cultures and contains 3 g of fiber (10% DV) and 40 mg of vitamin C (44% DV per fruit), making it a good source of fiber and an excellent source of vitamin C, in addition to a source of polyphenolic compounds.

Caraway

Caraway contains the monoterpene carvone; foods contain- ing carvone have a history of use as carminatives. Caraway essential oils inhibited colon carcinogenesis in rats. Carvone exhibited analgesic and anti-inflammatory properties in a laboratory study.

Carbohydrate

Macronutrient comprised of single monosaccharides (the simple sugars) or multiple monosaccharides (complex carbohydrate) that is metabolized aerobically or anaerobically for energy. Simple sugars are found naturally or may be added to fruit, milk, and grains. Complex carbohydrate fiber and starch is abundant in fruit, vegetables, grains, and legumes.

Carnitine

Also called l-carnitine which is the form found in food, as well as skeletal and cardiac muscle. Dietary sources of carnitine include meats, especially red meats; dairy products; breads; and vegetables. Carnitine transports long-chain fatty acids into the mitochondria for oxidation and energy production and transports waste compounds out of the mitochondria.

Carotenoids

Carotenoid sources include fruits, vegetables, and oils. Numerous observational studies have found that people who ingest more carotenoids in their diets have a reduced risk of several chronic diseases, including cancer, cardiovascular disease, age-related macular degeneration, and cataract.

Carrot

Root vegetable widely recognized for its role in vision because it is a rich source of provitamin A carotenoids. Vitamin A performs functions essen- tial to the eye and visual process (see also: Vitamin A). Carrots are a good source of insoluble fiber, and the purple cultivar is rich in anthocyanins.

Carrier wave

A wave that is modulated, or "tuned," to carry information. FM and AM are ways of modulating radio waves to send information

Chain reaction
A nucleus splitting atomic reaction that quickly snowballs. Used in nuclear reactors and atom bombs.

Catechin

Polyphenol flavonoids found in black, oolong, and green tea, apples, pears, chocolate, and broad beans. Types of catechins include epicatechin, epicatechin gallate, and epigallocatechin gallate (EGCG). A number of human observational studies found that tea catechins were associated with a reduced risk of stroke. A beneficial effect of a high intake of catechins is fighting against chronic obstructive pulmonary disease.

Cauliflower

Brassica vegetable consumed raw or cooked that is an excellent source of vitamins C and K and a source of glucosinolates. In a laboratory study, an antioxidant in cauliflower neutralized free radical activity and inhibited the peroxidation of linolenic acid.

Celery

While not an appreciable source of any particular nutrient, a stalk of celery contains about 2 g of fiber, and celery contains important flavonoids, such as the flavonol quercetin and flavones, such as luteolin and apigenin, and celery is considered to be a principal dietary source of flavones in the US diet, along with parsley and peppermint.

Chamomile

In a laboratory analysis, apigenin was anti-inflammatory, antioxidant, and anticarcinogenic properties.Chamomile has been traditionally and is currently used for sleeplessness, anxiety, and gastrointestinal conditions such as upset stomach, gas, and diarrhea, although its efficacy for any of these uses is unknown and studies examining chamomile alone are lacking.

Cherimoya

Cherimoya has been used in traditional Mexican medicine for its antianxiety, anticonvulsant, and tranquilizing properties. In an animal study, Annona cherimola extract administered intraperi- toneally significantly decreased plasma total cholesterol, triglycerides, and LDL cholesterol, and increased HDL cholesterol levels. Cherimoya is a source of fiber, magnesium, potassium, cryptoxanthin, lutein, and xeaxanthin.

Cherry, sweet

Smallest member of the Rosaceae stone-fruit family. Fresh, dried, or canned, 1/2 cup of raw sweet cherries contains vitamin C (5 mg 5.5% DV) and potassium (103 mg 5% DV) in addition to phenolic compounds that contribute to cherries' color, taste, aroma, and flavor, and cherries are a natural source of melatonin.

Chia

Chia seeds are a source of soluble and insoluble fiber (9.8 g total fiber, 35% DV per 1 oz) and supply linolenic acid. Other chia constituents include approximately 5 g of protein (10% DV) and 15.6 μg of selenium (28% DV) per 1 oz; in addition to quercetin, kaempferol, caffeic acid, and chlorogenic acid.

Chicory

In laboratory studies, Cichorium intybus exhibited antioxidant, antimicrobial, anthelmintic, antidiabetic, hepatoprotective, gastroprotective, analgesic, antiallergic, and tumor inhibitory activity.

Chive

One tablespoon (3 g) of fresh chives provides: 1 calorie; negligible fat, protein, and carbohydrate (less than 1 g of each), vitamins A (131 IU), C (1.7 mg), vitamin K (6.4 μg), folate (3 μg), calcium (3 mg), magnesium (1 mg), phos- phorus (2 mg); and potassium (9 mg). In a population study (n=238 men with confirmed cases of prostate cancer compared to n=471 male control subjects), intake of allium vegetables, including chives, was inversely associated with the risk of prostate cancer.

Chocolate

High in fat, 1 oz (28 g) of dark chocolate (45– 59% cacao solids) is a source of mag- nesium (41 mg, 9% DV) and a good source of iron (2.27 mg, 13% DV). Chocolate products and cocoa are "among the most concentrated sources of the procyanidin flavonoids catechin and epicatechin." Approximately 3.5 oz of dark chocolate (100 g) contains about 50 mg of flavanols and milk chocolate (100 g) contains about 13 mg of flavanols. Intake of flavonoid-rich foods and risk for cardiovascular disease are inversely related, due possibly to flavonoid induced changes in oxidant defense, vascular reactivity, and platelet reactivity.

Choline

Vitamin that serves as a precursor for the neurotransmitter acetylcholine, a methyl donor that is used for phospholipid synthesis, and is necessary to convert homocysteine to methionine. Preconception dietary intakes of choline between 350 and 544 mg or more were associated with reduced risk of neural tube defects in an epidemio- logical study.

Chromium

Ultra-trace mineral necessary for insulin action that is referred to as the
glucose tolerance factor. Chromium is widespread in plant and animal foods in minuscule amounts, and eating a variety of whole grains, fruits, vegetables, meats, and milk products is recommended to obtain adequate amounts.Chromium may improve insulin sensitivity, which can modify the risk of diabetes and cardiovascular disease (CVD).

Cilantro

Though normally consumed in insignificant quantities to be an appreciable source of nutrients, cilantro contains vitamins A and K, and contains numerous phytochemicals including caffeic acid, chlorogenic acid, quercetin, and limonene. Coriander has been used in traditional medicine to treat cystitis. Laboratory studies have shown coriander essential oils to have antioxidant and hepatoprotective properties. Coriander extract exerted antianxiety activity in a laboratory study.

Cinnamon

In animal models, cinnamon exhibited hypoglycemic, antimicrobial, antifungal, antiviral, antioxidant, antitumor, blood pressure-lowering, cholesterol lowering, lipid lowering, gastroprotective, and anticholinesterase properties.

Citrus

Juicy, segmented fruits of the genus Citrus that are vitamin C and folate rich and among the best sources of flavonoids, such as naringenin, hesperidin, nobiletin, and tangeretin, in addition to others. The skin, peel, and rinds of citrus fruits are rich in essential oils and contain more phytochemical compounds on a per gram basis than the edible interior flesh.

Clove

Clove oil, which contains eugenol, is most popularly known as being a toothache remedy topically; however, according to FDA, efficacy is lacking for this use. Clove was shown to have anti-inflammatory effects in a laboratory study. Clove is not effective for vomiting, upset stomach, nausea, gas, or diarrhea.

Cloudberry

Cloudberry is a nutritious fruit containing vitamins C, A, and E, along with antioxidants and dietary fiber. Scientific research highlights its potential health benefits, including antioxidant effects, anti-inflammatory properties, and support for digestive health. Including cloudberry in your diet or skincare regimen may offer various health advantages, particularly in regions where it is native and traditionally used.

Coconut

The flesh of coconuts is a source of potassium and saturated fat, in addition to flavonoids and saponin. Coconut water contains potassium (400 mg, 8% DV per 1 cup). Diets rich in coconut oils have been shown to reduce coronary artery disease risk factors, such as tissue plasminogen activator antigen and lipoprotein(a); however, current recommendations are to limit saturated fats, including coconut oil to no more than 7% of calories.

Coenzyme

A fat-soluble antioxidant synthesized in the body168 10 that occurs in virtually all cells—it is ubiquitous—hence it is also known as "ubiquinone." A participant in ATP generation in aerobic metabolism that is essential for electron and proton transport in the mitochondrial respiratory chain. It is present in the highest quantities in the heart, liver, kidney, and pancreas. Endogenous coenzyme Q10 production decreases with age. Secondary deficiency may be linked to the use of statins to treat hyperlipidemia.

Coffee

Its caffeine constituent prevents fatigue and improves mental alertness, varying in content per 1 cup (8 oz) between 75–150 mg with darker roasts providing less caffeine than light roasts.

Copper

Trace mineral found in shellfish, nuts, beans, organ meats, and whole grains that functions as a component of a number of metalloenzymes and that is involved in red blood cell formation, immunity, and in mainte- nance of blood vessels, nerves, and bones. Copper deficiency, although rare, causes normocytic, hypochromic anemia, leukopenia, and neutropenia. Copper is an antimicrobial agent.

Corn

Corn is a source of fiber (1.8 g, 5% DV per 1 cob) and phytochemicals such as lutein, xeaxanthin, and ferulic acid. It is notably deficient in the amino acids lysine and tryptophan, the latter of which is converted into niacin.

Crappie

While crappie provides vitamins B12, D, A, and E in modest amounts, its main nutritional benefits come from high-quality protein, omega-3 fatty acids, and essential minerals. Including crappie as part of a balanced diet can contribute to overall health, particularly benefiting heart health and providing essential nutrients.

Cranberry

Cranberry is a source of choline, vitamins C, A, and K,194 and proanthocyanidins. Daily doses of 120–4,000 mL/day of cranberry juice or 400 mg of cranberry extract have been used to help prevent urinary tract infections (UTIs) based on the results of small studies that postulated cranberry prevents bacteria from sticking to the cells that line the bladder.

Cucumber

Cucumbers contain vitamin K (8.5 μg, 6% DV per 1/2 cup) and are so low in calories (8 calories per 1/2 cup) that they are considered to be calorie-free for purposes of weight and diabetes management. Cucurbitacins demonstrated anticancer properties in colon, breast, lung, and central nervous system cancer cell lines, and inhibited the COX-2 enzyme and lipid peroxidation.

Cumin

It is usually consumed in minuscule amounts, cumin's nutrient content is negligible but includes antioxidant vitamins. Cumin is used in traditional Indian medicine for the treatment and management of sleep disorders, indigestion, and hypertension. Cuminum cyminum in animal models exerted antihypertensive, antihyperglycemic, and anticarcinogenic effects.

Top 9 Foods to Help Increase Blood Flow:

Pomegranate juice
Pomegranate juice is rich in antioxidants, which help open the blood vessels to allow more blood to flow through. It is also known for keeping the arteries from becoming stiff and thick.

1. Onions
   Onions are an excellent source of flavonoid antioxidants, which benefit heart health and circulation by helping your arteries and veins widen when blood flow increases.

2. Fatty fish
   A diet rich in fish that are low in saturated fat and high in Omega-3 fatty acids has been associated with improved blood flow and a reduced risk of heart attack and stroke.

3. Beets
   Beets are rich in natural chemicals called nitrates. Through a chain reaction, the body converts these nitrates into nitric oxide, a chemical that helps improve blood flow and blood pressure.

4. Leafy greens
   Leafy greens thin the blood and help oxygen circulate around the body more efficiently.

5. Citrus fruits
   Consuming flavonoid-rich citrus fruits like oranges, lemons and grapefruit may decrease inflammation in your body, which can reduce blood pressure and stiffness in your arteries while improving blood flow. Watermelon can also improve blood flow, as it contains lycopene, a natural antioxidant linked to improving circulation.

6. Walnuts
   Nuts like walnuts and almonds help improve circulation by reducing inflammation and oxidative damage in the arteries.

7. Tomatoes
   The lycopene in tomatoes can protect against cardiovascular disease. The vitamin K in tomatoes helps control bleeding and blood clotting as well as improves circulation.

8. Berries
   Blueberries and strawberries contain flavonoids, which help dilate arteries, reduce plaque buildup and increase blood flow.

Carbohydrate
Any of a variety of compounds that are made of carbon and water.

Carbon
The sixth element on the periodic table with 6 protons, 6 neutrons, and 6 electrons. It has the symbol C.

Carboxylic acid group
A carbon, hydrogen, and two oxen atoms.

Cellulose
The polysaccharide that is the main structural molecule in plats.

Chemistry
The field of science that studies that composition, structure and properties of matter.

Chemical elements
All of the atoms that make up living and non-living things.

Chromatography
A method for separating mixtures by passing a solution over a

solid film or powder.

Citric
An acid found in various fruits including grapefruit and oranges.

Chemical reaction
Occurs when bonds between atoms and molecules are created or destroyed.

Combination reaction
When two or more molecules combine with each other to make a new molecule.

Concentration
The number of molecules in a given volume of solution.

Covalent bond
Attachment formed when two atoms are joined together and share electrons.

Cross-link
The links between two polymer molecules.

Cytosine
One of the four nucleic acid bases that make DNA.

Chemosynthesis
The ability of living things to make the essential chemicals for life using energy from chemical reactions.

Chloroplasts
Tiny little organelles inside a plant cell that harvest energy from sunlight. Originally they were photosynthesizing cyanobacteria.

Contagion

The spread of a disease from one person to another. A disease that is easily transmitted is contagious.

Contamination
The addition of a pathogen or poison to make something unhealthy.

Cytoplasm
The watery fluid that fills a cell.

Cellulose
A material made from the cell walls of green plants and useful in making paper and fabrics; often obtained from wood pulp or cotton.

Dandelion

Dandelion contains calcium (103 mg, 8% DV) and is an excellent source of vitamins C (19 mg, 25% DV) and K (428 μg, 356% DV). It also supplies phytochemicals including taraxasterol. The root is a source of triterpenes, steroids, and inulin. The study found significant increases in the frequency of urination in the 5-hour period after the first dose, and there was a significant increase in the excretion ratio (the ratio of urination volume to fluid intake) in the 5-hour period after the second dose. A third dose failed to change any measured parameter.

Dashi

Dashi is a flavorful Japanese soup stock primarily valued for its umami taste rather than its vitamin content. It provides essential nutrients from its ingredients like iodine from kombu and small amounts of B vitamins from bonito flakes and shiitake mushrooms. Incorporating dashi into dishes can enhance flavor and provide some nutritional benefits, especially when used in a balanced diet with other nutrient-rich foods.

Date

Two Medjool dates, a common type eaten as a snack and used for baking, contain potassium (280 mg, 6% DV), while 1/4 cup is an excellent source of potassium (1124 mg, 33% DV). Dates have the highest total polyphenol content among commonly eaten fruits.

Dill

Dill is a widely used traditional medicinal plant used to treat various ailments, such as the prevention and treatment of gastrointestinal (e.g., indigestion, flatulence, colic pain), kidney, and urinary tract diseases and disorders; sleep disorders; and spasms. Dill killed bacteria that cause nosocomial infection, in particular, P. aeruginosa, and important pathogens associated with indigestion, dysentery, and diarrhea in laboratory tests. In laboratory studies, dill was shown to be anti fungal and effective against vulvovaginal candidiasis in immunosuppressed mice.

Docosahexaenoic Acid

Docosahexaenoic acid (DHA) is an omega-3 polyunsaturated fatty acid crucial for its anti-inflammatory properties, immune modulation, and roles in reducing serum triglycerides and cholesterol. It is synthesized in the body from alpha-linolenic acid to eicosapentaenoic acid (EPA) and then to DHA, primarily found in fatty fish like salmon and fortified foods such as omega-3-fatty-acid-fortified eggs. DHA is vital for neurocognitive development, abundant in the brain, retina, and spermatozoa, and deficiency has been linked to cognitive decline, neurodegenerative diseases, and conditions like ADHD in children.

Decomposition reaction Occurs when a molecule breaks apart into two or more molecules.

Deoxyribonucleic acid a polymer of A, T, G, and C that carries the genetic code.

Dilute
A solution with few molecules per unit volume.

Displacement reaction occurs when one atoms displaces another atom during a chemical reaction.

Disole
Occurs when a liquid like water causes the molecules in a solid to break apart from each other and become loose in solution.

Double helix
The structure formed by two strands of DNA wrapped around each other.

DNA
Deoxyribonucleic acid.

DNA polymerase
The protein that copies DNA inside a cell.

Digital
When an electric signal is a sequence of binary numbers. Each number is equal to the size of the voice or picture being transmitted at that moment.

Dynamo
Another word for a generator a machine that turns motion into electrical energy.

Doping
Adding impurities to an electrical nonconducting material to make it a semiconductor.

Denature

To destroy the properties of a biological chemical that make it useful, usually by heat or acidity.

Eel

Eel provides vitamins A, B12, D, and E, along with omega-3 fatty acids and essential minerals. It offers several health benefits, including support for heart health, brain function, and overall well-being. Incorporating eel into a balanced diet can contribute to meeting daily vitamin and nutrient needs while enjoying its culinary versatility and nutritional benefits.

Eggplant

Eggplant is a good source of fiber, and contains anthocyanins, phenolics, saponins, terpenoids, and steroidal alkaloids. While eggplant has not been shown to reduce blood lipids, an experimental study showed phenolic-enriched extracts of eggplant to have moderate to high angio- tensin I-converting enzyme inhibitory activity, suggestive of potential antihypertensive properties.

Eicosapentaenoic Acid

Eicosapentaenoic acid (EPA) is one of the predominant omega-3 fatty acids, synthesized in limited amounts in the body from alpha-linolenic acid along with DHA. EPA is essential for growth and cognition, found in significant quantities in fatty fish such as mullet, mackerel, salmon, and others like anchovy, sardines, and tuna. Human milk also contains EPA. Current average intake of EPA and DHA in the US is around 100 mg/day, which experts consider suboptimal. Alterations in essential fatty acids (EFAs) are associated with conditions like obesity, hypertension, diabetes mellitus, coronary heart disease, alcoholism, schizophrenia, Alzheimer's disease, atherosclerosis, and cancer, though whether these alterations are causes or effects of these conditions remains unclear.

Endive

Endive contains folate (35 μg, 8% DV) and is a good source of vitamin A (540 IU, 18% DV) and an excellent source of vitamin K (55 μg, 46% DV) (values per 1/2 cup). Kaempferol is one of many phytochemicals in endive, and a red cultivar of endive (Cichorium intybus L. cultivar) is a source of anthocyanins. In laboratory studies, Cichorium intybus inhibited free radical-mediated DNA damage, and exerted cytoprotective and antiproliferative effects in cells.

Elderberry

Elderberries contain vitamin C and are rich in antioxidants, particularly anthocyanins, which contribute to their potential health benefits. Scientific findings support their use for immune support, anti-inflammatory effects, and potential antiviral activity against influenza viruses. Incorporating elderberries into your diet or using elderberry supplements may provide additional health advantages, particularly during cold and flu seasons.

Elk

Elk meat is a nutritious protein source rich in vitamins B12, B6, niacin, and riboflavin, along with beneficial omega-3 fatty acids and essential minerals like iron and zinc. Scientific findings support its role in providing lean protein, supporting heart health, and contributing to a balanced diet. Incorporating elk meat into meals can offer a variety of health benefits, particularly for those seeking a leaner alternative to traditional meats.

Ethanol

Ethanol, also known as ethyl alcohol or simply alcohol, is

a molecule produced during the anaerobic fermentation of sugars found in berries, grains, and other carbohydrates. It is a psychoactive compound and central nervous system depressant. Ethanol provides approximately 7 kilocalories per gram and is a significant calorie source in alcoholic beverages like beer, wine, and distilled liquor.

9 Best Foods For Eye Health

1. Fish

Fish are rich in omega-3 fatty acids, which provide your body with a range of benefits. Most importantly for us is the benefits that it can have on our eye health.Some of the best fish that are richest in omega-3 include tuna, salmon, mackerel, sardines, and anchovies.

2. Raw red peppers

Peppers are one of the vegetables that contain the most vitamin C per calorie. Vitamin C is great for the blood vessels in your eyes and can lower your risk of getting cataracts.Although you can get vitamin C supplements, it's just as easy to get it in a range of different foods like strawberries, cauliflower and bok choy.

3. Seeds

Seeds are also very high in Omega-3s and vitamin E which are both beneficial for eye health.Choose seeds like chia seeds, flax seeds, or hemp seeds for the optimal amount of fatty acids and vitamins.

4. Dark, leafy vegetables

Dark, leafy vegetables like spinach and kale are great foods to eat to fuel your body and improve your overall health. These foods are particularly rich in antioxidants called carotenoids which are thought to reduce the risk of developing or slowing AMD (age-related macular degeneration) the most frequent cause of severe visual loss in those over the age of 55 years.They also contain lutein and zeaxanthin which are great sources of vitamin C.

5. Carrots

Although they may not be able to make you see in the pitch-black, orange fruit and vegetables like carrots, sweet potatoes and mangoes are high in beta-carotene.This is a type of vitamin A that can help your eyes adjust to see in the dark better – so there is truth in what we were all told as children.

6. Lean meat and poultry

Anything that contains high levels of Zinc is great for our diet as it brings vitamin A from your liver to your retina. Once there, it's used to make melanin which can help to protect your eyes as well as your skin.Although oysters have more zinc than any other foods, you can also eat beef, pork, and chicken to increase your intake of eye-friendly minerals.

7. Eggs

Like lean meats, eggs also contain a lot of zinc that helps your body utilise the lutein and zeaxanthin that are found in egg yolk.These organic compounds help your eyes block the harmful blue light that could potentially cause damage to your retina.

8. Broccoli and Brussels sprouts

They might not be everyone's favourite vegetables, but you can't argue when it comes to their nutritional benefits.Broccoli and Brussels sprouts contain a combination of vitamin A, vitamin C, and vitamin E that help protect your eyes from free radicals.

9. Beans and legumes

If you're on a vegan or vegetarian diet, the best foods to eat for your eye health include beans and legumes.These low-fat, high-fibre foods are high in zinc and various other vitamins that can help keep your vision sharper for longer.

Basic nutritional elements for eye health

Here is a breakdown of the actual nutritional elements that are key when optimising your diet for eye health: If you want to maintain good eye health:

- vitamin C
- vitamin E
- lutein
- zeaxanthin
- zinc oxide
- copper oxide

By incorporating these nutritional elements into your diet regularly, you're ensuring that your eyes are getting everything they need.

Top 10 foods with emollients

1. Avocado: Avocado is rich in healthy fats and oils, such as oleic acid, which can soften and moisturize the skin.
2. Olive Oil: Olive oil contains oleic acid and other fatty acids that can hydrate and nourish the skin, making it smoother.
3. Coconut Oil: Coconut oil is rich in medium-chain fatty acids that can penetrate the skin and provide emollient benefits.
4. Shea Butter: Shea butter is known for its high concentration of fatty acids and vitamins, making it an excellent emollient for dry skin.
5. Cocoa Butter: Cocoa butter is derived from cocoa beans and is high in fatty acids that can hydrate and soften the skin.
6. Almond Oil: Almond oil is rich in vitamin E and fatty acids, which can moisturize and improve the skin's barrier

function.

7. Jojoba Oil: Jojoba oil is similar to the natural oils produced by the skin and can help balance oil production while providing emollient benefits.

8. Sunflower Seed Oil: Sunflower seed oil is lightweight and rich in linoleic acid, which can help maintain the skin's barrier and prevent moisture loss.

9. Grapeseed Oil: Grapeseed oil is light and easily absorbed, containing antioxidants and fatty acids that can nourish and moisturize the skin.

10. Soybean Oil: Soybean oil is rich in linoleic acid and other fatty acids that can improve skin hydration and smoothness.

Electrode
The piece of a pH meter that is placed in the solution for detecting the pH, sometimes made of glass.

Electron
On of the three fundamental particles that make atoms. It has almost no mass compared to protons and neutrons, and it carries a negative charge.

Electron cloud
The space surrounding the nucleus of an atom that houses the electrons.

Element
One of more than 100 fundamental units called atoms.

Evaporation
Occurs when a liquid, such as water, turns into a gas and escapes into the air. Can be used to separate mixtures where one

substances will evaporate and the other will not.

Exchange reaction
Occurs when one atoms trades places with another atom during a chemical reaction.

Enzyme
A nonliving chemical that helps control the rate of processes inside cells and the bodies of living things.

Eukaryote
A living thing whose cells store their DNA in a central nucleus. Eukaryotes include single celled and multicellular organisms.

Electric charge
A property of matter that makes charged particles feel a force when they come close to one another. The two types of electric charge positive and negative make electricity possible.

Electromagnetism
A fundamental force of nature between electrically charged particles.

Electron
A tiny, subatomic, negatively charged particle that orbits around the nucleus of an atom.

Element
A chemically pure substance consisting of atoms of only one type.

Fatty Fish
Fatty fish are high in protein, vitamin D, and omega-3-fatty acids, and are sources of coenzyme Q10. Fish oil constituents DHA, EPA, and vitamin D are anti-inflammatory. People who eat seafood one to four times a week are less likely to die of heart

disease. Epidemiological studies show that the dietary intake of n-3 PUFAs through fish consumption is inversely correlated with the prevalence of depression.

Fennel

Fennel experimentally improved hypertension and glaucoma, and anethole demonstrated antiplatelet aggregatory properties and provided "significant protection" from ethanol-induced gastric lesions in animal models. Clinical trial data suggests fennel extracts may have potential in treating infantile colic.

Fenugreek

Fenugreek reduced serum glucose in diabetes experimentally and in a few small clinical trials. The gum within the fenugreek seed fiber reduces serum glucose and cholesterol. There is not enough evidence to support its use as a galactagogue or a pregnancy inducer. Fenugreek fiber significantly increased satiety in a small, single-blind, randomized trial of healthy obese patients (n = 18).

Ferulic Acid

Phytochemical found in seeds and leaves made from the metabolism of the amino acids phenylalanine and tyrosine. Found in high levels in vegetables, fruits, cereals, and coffee with the average intake estimated to be 150–250 mg/day. In laboratory studies, ferulic acid exhibited antioxidant, antimicrobial, anti-inflammatory, antithrombotic, anticancer, and increased sperm viability effects.

Fiber

There are two main types of dietary fiber: soluble and insoluble.Soluble Fiber, found in foods like citrus fruits, apples, bananas, legumes, oats, and barley. It dissolves in water and forms a gel-like substance in your gut. Benefits ; Slows digestion, helps lower cholesterol, and stabilizes blood sugar levels. Insoluble Fiber ; Found in whole grains, wheat bran, vegetables like cabbage and carrots, and fruit skins. It adds bulk to stools and supports regular bowel movements.

Benefits ; Prevents constipation, aids in digestion, and promotes a feeling of fullness.Eating foods rich in fiber supports heart health, regulates blood sugar, improves digestion, and helps with weight management.

Fiddlehead

Fiddleheads are rich in vitamins A, C, and K, along with antioxidants and dietary fiber. Scientific findings highlight their potential health benefits, including antioxidant and anti-inflammatory properties. Incorporating fiddleheads into a balanced diet can provide essential vitamins and contribute to overall health, especially when prepared and consumed safely.

Fig

It is mild in flavor and a good source of fiber (1.4 g, 10% DV per fig). Phytochemical components include phenolics, coumarins, flavonoids (e.g., anthocyanins, quercetin, luteolin), and terpenoids. Figs demonstrated antioxidant properties and suppressed various types of cancer cells in vitro.

Flathead Catfish

Flathead catfish provides vitamins B12, D, A, and E, along with omega-3 fatty acids and essential minerals. It offers several health benefits, including support for heart health, brain function, and overall well-being. Including flathead catfish in a balanced diet can contribute to meeting daily vitamin and nutrient needs while benefiting from its lean protein and omega-3 content.

Flavonoids

Flavonoids are diverse plant chemicals common in the American diet. They come in groups like anthocyanidins, catechins, flavones, and more, giving foods their colors and flavors. Found in foods like fruits, vegetables, nuts, tea, and wine, they've been used in traditional medicine for ages. Despite being discovered in the 1930s, they're the main type of plant chemicals in our diets. Daily intake varies widely, but they're linked to heart health and possibly fighting cancer. They might also help with HIV by blocking reverse transcriptase. Overall, flavonoids play a big role in health and nutrition.

Flaxseed

Flaxseed are 35–45% fiber, making them an excellent source of fiber of which 2/3 is insoluble and 1/3 is soluble.65 Flaxseeds contain potassium (341 mg, 7% DV per 1/4 cup) are a good source of iron (2.4 mg of iron, 13% DV), in addition to being a rich source of linolenic acid and beta-sitosterol. Flaxseed oil suppresses oxygen radical production by white blood cells, prolongs bleeding time, and in higher doses suppresses serum levels of inflammatory mediators, but is not an effective hypolipemic agent.

Folate

Folate, also called folic acid, is a vital vitamin essential for DNA synthesis and cell growth. It needs vitamin B12 to become active in making DNA. Synthetic folic acid in fortified grains and supplements is easier for the body to absorb than natural folate in foods. Around 8% of Americans don't get enough folate based on what's recommended. Normal levels of folate in the blood range from 3 to 16 ng/mL, and in red blood cells from 140 to 628 ng/mL. Not having enough folate can lead to megaloblastic anemia and problems with the digestive system because cells can't divide or make proteins properly. Low folate levels, along with low levels of vitamins B6 and B12, can cause high homocysteine levels (hyperhomocysteinemia), which raises the risk of various health problems and death. To prevent birth defects in babies, women who might get pregnant should have 400 μg of folate or folic acid daily before pregnancy and 600 μg during pregnancy. Studies suggest that more folate in the diet might lower the risk of pancreatic and maybe colon cancer. Having enough folate is also linked to a lower risk of a certain type of stroke. Low folate levels are found in people with mood problems who take lithium and in people with alcoholism and depression, which might affect how well they respond to antidepressants.

Fructan

Also called fructooligosaccharide (FOS). Naturally occurring polymer of fructose, an example of which is inulin. Fructans are natural sweeteners that have a low caloric value, do not lead to a rise in serum glucose, do not stimulate insulin secretion, promote the growth of intestinal bifidobacteria, and may improve the absorption of certain minerals. A review of 34 studies including experimental and human clinical trials concluded that FOS may have a beneficial effect on lipid metabolism and regulation of serum cholesterol levels when combined with lifestyle and dietary changes.

Fruit

Fruits, often overlooked, are rich sources of essential nutrients like vitamins A and C, folate, potassium, and fiber. They also provide various phytochemicals that vary by color (pigment):

- Blue/purple plant foods contain anthocyanidins, flavonols, flavan-3-ols, proanthocyanidins, ellagic acid, and resveratrol.

- Green plant foods typically contain flavones, flavanones, flavonols, beta-carotene, lutein, xeaxanthin, indoles, isothiocyanates, and organosulfur compounds.

- White plant foods typically contain flavonols, flavanones, indoles, isothiocyanates, and organosulfur compounds.

- Yellow plant foods typically contain flavonols, flavanones, alpha-carotene, beta-carotene, beta-cryptoxanthin, and xeaxanthin.

- Red plant foods typically contain anthocyanins, flavonols, flavones, flavan-3-ols, flavanones, proanthocyanidins, lycopene, ellagic acid, and resveratrol.

Diets rich in plant foods, including fruits, are associated with a lower risk of cancers affecting the mouth, pharynx, larynx, esophagus, stomach, and lung. There's also evidence suggesting these diets reduce the risk of colon, pancreas, and prostate cancers. Moreover, such diets lower the risk of diabetes, heart disease, and hypertension, help control calorie intake, and aid in weight management. They are also linked to a reduced incidence of age-related diseases such as Alzheimer’s Disease.

Foods That Help You Burn Fat (That you may be eating already) ...

1. Fatty fish
Salmon, herring, sardines, mackerel, and other fatty fish contain omega-3 fatty acids, which may help you lose body fat.

Fish is also an excellent source of high quality protein, which may lead to greater feelings of fullness and can help increase your metabolic rate.

2. MCT oil

Medium-chain triglycerides (MCT) oil is made by extracting MCTs from palm oil. It's a type of fat that's metabolized differently than the long-chain fatty acids found in most foods.

Research suggests that MCTs may help increase metabolic rate, reduce hunger , and promote better retention of muscle mass during weight loss.

Studies suggest that replacing fat in your diet with – 2 tablespoons (tbsp) of MCT oil per day can boost fat burning. Try to start with 1 teaspoon (tsp) and gradually increase to avoid digestive side effects.

3. Coffee

Coffee is one of the most popular beverages worldwide.

Coffee a great source of caffeine, which may help you burn fat. To get the fat-burning benefits of caffeine without the potential side effects, such as anxiousness or insomnia, aim for no more than 400 mg per day. This is the amount found in about 4–5 cups of coffee, depending on its strength. Some people are more sensitive to caffeine's effects and may need to limit their intake.

4. Eggs

Although egg yolks have traditionally been avoided due to their high cholesterol content, whole eggs may help with weight loss.

For one, they are very nutrient-dense and contain a lot of protein, which can help ward off hunger and overeating. For example, a 2017 study found that eating a high-protein diet reduced the subject's hunger by 16% while increasing daily fullness by 25%.

One of the reasons eggs are so filling may also be due to the boost in calorie burning that occurs during protein digestion.

Eating up to three eggs a week can help you burn fat while keeping you full and satisfied. More than this amount has been associated with a higher chance of heart disease.

5. Green tea

In addition to providing a moderate amount of caffeine, green tea is an excellent source of epigallocatechin gallate (EGCG), an antioxidant that promotes fat burning and the loss of belly fat.

Though research suggests that drinking green tea may help improve your metabolism and lower your body fat, more research is necessary to support these claims.

That said, drinking about 2-3 cups of green tea daily may be optimal for providing the variety of health benefits.

6. Whey protein

Consuming whey protein may help you suppress appetite, especially because it stimulates the release of "fullness hormones," such as PYY and GLP-1 (almost as good as fiber).

Moreover, whey appears to boost fat burning and promote weight loss. For this reason, a whey protein shake is a quick meal or snack option that promotes fat loss and may help improve your body composition.

7. Apple cider vinegar

Apple cider vinegar is an ancient folk remedy with evidence-based health benefits.

Animal research suggests that vinegar's main component, acetic acid, may help increase fat burning and reduce belly fat storage in several animal studies. However, more human studies are needed to verify this.

Start with 1 tsp per day diluted in water and gradually work up to 1 tbsp per day to minimize potential digestive discomfort.

8. Chili peppers

One of these is called capsaicin, and consuming it may help you achieve and maintain a healthy weight by promoting fullness and preventing overeating.

Consider eating chili peppers or using powdered cayenne pepper to spice up your meals several times a week.

9. Oolong tea

Oolong tea is contains polyphenols, which are compounds associated with helping reduce things like blood sugar and and body weight

Like other teas, it also contains caffeine, which helps weight and body fat loss.

Drinking a few cups of green tea, oolong tea, or a combination of the two on a regular basis may promote fat loss and provide other beneficial health effects.

That said, most research on oolong tea and weight loss is based on animals, so more human studies are needed.

10. Full-fat Greek yogurt

Full-fat Greek yogurt is extremely nutritious. First, it's an excellent source of protein, potassium, and calcium.

Research also suggests that eating high protein dairy products can boost weight and fat loss.

Eating 2 servings of dairy such as Greek yogurt daily may provide a number of health benefits. But make sure to choose plain, full-fat Greek yogurt.

11. Olive oil

Olive oil is one of the healthiest fats on earth.

Most of olive oil is composed of oleic acid, which has been shown to have a positive effect on fat and body mass.

To incorporate olive oil into your daily diet, drizzle a couple of tablespoons on your salad or add it to cooked food.

Top 10 Sources of Fiber Rich Foods

1. Beans

Lentils and other beans are an easy way to sneak fiber into your diet in soups, stews and salads. Some beans, like edamame (which is a steamed soy bean), are even a great fiber-filled snack. There are 9 grams of fiber in a half cup serving of shelled edamame. A bonus? All of these provide a source of plant protein, too. Some bakers have even started including beans or bean flours in their baked goods, which research proves can still make quality cakes.

2. Broccoli

This veggie can get pigeonholed as the fiber vegetable. Its cruciferous nature meaning it's from the Brassica genus of plants along with cauliflower, cabbage and kale makes it rich in many nutrients in addition to fiber. Studies have shown that broccoli's five grams of fiber per cup can positively support the bacteria in the gut, which may help your gut stay healthy and balanced.

3. Berries

Berries get a lot of attention for their antioxidants, but they're full of fiber, too. Just a cup of fresh blueberries can give you almost four grams of fiber, and there is nearly the same amount of fiber in a cup of frozen unsweetened blueberries. Blackberries, strawberries and raspberries are also great sources of fiber. Of course, one of the biggest benefits of berries is that they're naturally low in calories, too.

4. Avocados

Avocados pretty much go with everything—toast, salads, entrees, eggs—and while they're often recognized for their hefty dose of healthy fats, there are 10 grams of fiber in one cup of avocado (so just imagine how much is in your guacamole).

5. Popcorn

There's one gram of fiber in one cup of popcorn, and the snack (when natural and not covered in butter, like at the movies) is a whole grain that can satiate cravings with a hit of fiber. It's even been called the King of Snack Foods.

6. Whole Grains

Good news for bread lovers: Real whole grains, found in 100% whole wheat bread, whole wheat pasta, brown rice, and oats, have fiber. One tip to watch out for: as required by The Food and Drug Administration, whole grains should be the first ingredient on a food package in order for it to be considered a real whole grain.

7. Apples

That old saying that "an apple a day keeps the doctor away" isn't necessarily true, according to research, but the fruit can boost your fiber intake. There are about four grams of fiber in an apple, depending on its size. And, of course, they're a nice and crunchy snack.

8. Dried Fruits

Dried fruits like figs, prunes and dates can boost your fiber intake dramatically and are recommended for those struggling with constipation. The sugar called sorbitol, which naturally occurs in these fruits, can help your bowels and lead to more comfort. However, eating too many can lead to cramping or diarrhea, so try a small serving and see how you feel once you've digested them, before noshing on too many more.

9. Potatoes

Sweet potatoes, red potatoes, purple potatoes and even the plain old white potato are all good sources of fiber; one small potato with skin can provide close to three grams of fiber. The veggie has a bad reputation for running in the wrong crowds—fries and chips, to name a few. However, when not fried in oil and slathered in salt, potatoes can provide many benefits.

10. Nuts

Nuts aren't just a great source of protein and healthy fats—sunflower seeds and almonds each have more than three grams of fiber in a serving. They can help you reach the 25-gram intake of fiber recommended by the FDA for women and 38-gram recommendation for men. Raw or dry-roasted nuts are preferred over the pre-packaged variety (which are usually cooked in oils that can add extra, unnecessary calories. Even nut butters can pack a punch of fiber.

Top 10 filaggrin rich foods

Oats: Oats contain beta-glucans, which are polysaccharides that can help soothe and moisturize the skin. They also support overall skin health.

Salmon: Rich in omega-3 fatty acids, salmon supports

skin barrier function and reduces inflammation, which can indirectly support filaggrin production.

Eggs: Eggs are a good source of biotin, which is important for skin health and may support the production of filaggrin.

Sweet Potatoes: Rich in beta-carotene, sweet potatoes can be converted into vitamin A in the body, which is essential for skin health and maintenance.

Spinach: Spinach is packed with vitamins A, C, and E, as well as antioxidants, which support overall skin health and function.

Almonds: Almonds are rich in vitamin E and healthy fats, which support skin hydration and may contribute to overall skin barrier health.

Avocado: Avocado is rich in healthy fats and vitamin E, both of which are important for skin health and barrier function.

Broccoli: Broccoli is high in vitamins C and K, which contribute to collagen synthesis and overall skin health.

Berries: Berries are rich in antioxidants such as vitamin C, which support collagen production and overall skin health.

Sunflower Seeds: Sunflower seeds are high in vitamin E and other nutrients that support skin barrier function and hydration.

Field of force
The area where on object feels a force from another object.

Fission
Splitting the nuclei of heavy atoms.

Fundamental force
One of four crucial interactions that govern how the universe works. Electromagnetism and gravity are long range; the strong and weak forces are short range.

Fundamental particle
The most basic part of matter a particle that can't be split into smaller, simpler parts.

Fusion
Joining together the nuclei of light atoms.

Fungi
A major kingdom of living spore producing organisms. The kingdom includes molds, yeast, and mushrooms.

Frequency
The rate at which something repeats itself the number of cycles it makes per unit of time.

Friction
A force that slows things down or resists forward movement, often by surfaces rubbing together; also called drag.

Garbanzo bean/Chickpea

A good source of fiber (3 g, 11% DV per 1/2 cup), garbanzo beans also contain numerous phytochemicals, including flavonoid glycosides. Garbanzo beans have a low glycemic index and may be used as a low- fat substitute for meat. One oz of meat, poultry, or fish is nutritionally equivalent to 1/4 cup cooked of garbanzo beans, although the iron content and absorption is lower in garbanzo beans, as is the case for all legumes, compared to meat.

Garlic

When crushed, it releases compounds like allicin and diallyl trisulfide, which may have health benefits. Garlic also contains flavonoids and terpenoids.

Historically, garlic was valued for allicin, known for its flavor and scent, but it's unstable and quickly degrades. The health effects of garlic likely come from its organosulfur compounds.

Studies initially suggested garlic might lower cholesterol and slow artery disease, but trials using fresh garlic, powdered garlic, or aged extracts showed inconsistent results. Methyl allyl trisulfide, another compound, may help prevent blood clots. Garlic has been studied for reducing heart arrhythmias and blood pressure in hypertensive people, but evidence is mixed. Recommended doses vary: 4 grams of fresh garlic daily for cholesterol and 600-900 mg of garlic powder for hypertension. Regarding cancer prevention, garlic's effects are unclear. While lab studies hint at anticancer properties, clinical evidence linking garlic to lower cancer risks remains inconclusive. Garlic also shows antifungal properties in labs, especially aged extracts, which also have antioxidant benefits and protect the liver and brain. Allicin and its derivatives have antimicrobial properties and have been studied for potential use against certain cancers.

In conclusion, while garlic is popular and shows promise for health benefits, more research is needed to confirm its effects, especially in preventing diseases like cancer and heart disease.

Ginger

Ginger contains niacin and phytochemicals including

phytosterols, berberine, and the pungent compound zingerone. Ginger has been used in traditional Asian medicine to alleviate stomachache, nausea, and diarrhea. Ginger promotes saliva production and gastric juice secretion, and pro- duces an increase in the tone and peristalsis of the intestine. Ginger may help to reduce nausea due to cancer chemotherapy when used in addition to conventional anti-nausea medication.

Glucosinolates

Secondary plant metabolites found in Brassica plants that transform into various active compounds including sulforaphane and indole-3-carbinol. Sprouts of Brassica vegetables typically contain significantly higher concentration of glucosinolates than mature plants.

Glutathione

Compound that is synthesized in the body and is a component of the antioxidant enzyme glutathione peroxidase. Fresh fruits, vegetables, and meats have moderate to high amounts of glutathione. The spice cumin is also a source of glutathione. Glutathione is thought to prevent free radical formation in humans.

Glycosides

Glycosides were shown to be antioxidants, suppress cancer cell growth, and exert antiatherogenic properties in laboratory studies.

Grains

Seeds known as grains are found atop stems, especially in the grass family. Cereals are grains where the outer husk (bran) covers the central endosperm and inner germ. The bran is fiber-rich with phytates, while the endosperm is starch-heavy, and the germ contains fats, proteins, and vitamins like vitamin E. Processing (refining) removes bran and germ for longer shelf life and taste, depleting nutrients such as fiber, fats, proteins, vitamins, and phytochemicals. "Enriched" refined grains have added thiamin, riboflavin, niacin, folic acid, and iron.

Studies link non-whole grain intake to higher pancreatic adenocarcinoma risk but show a lower risk with whole grain consumption.

Grains,Whole

Whole grains are grains that include the entire kernel bran, germ, and endosperm providing fiber, B vitamins, vitamin E, and various phytochemicals like antioxidants. Examples are oats, popcorn, and brown rice, or products labeled "whole" in their ingredients, such as "Whole wheat.” Research suggests that diets rich in whole grains may reduce the risk of cancer, heart disease, diabetes, and obesity. For instance, they may lower colorectal cancer risk by diluting carcinogens, speeding digestion, and producing beneficial fatty acids. Similarly, higher intake is linked to lower pancreatic cancer risk, possibly by improving insulin and lipid levels. However, clinical trials have not consistently shown significant cardiovascular benefits from whole grains, nor strong preventive effects against Type II diabetes. Still, diets emphasizing whole grains have helped with weight loss and improved blood pressure in overweight individuals.

Grape

Marble-sized snack fruit that contains numerous

phytochemicals, depend- ing on variety, such as catechin, epicatechin, resveratrol, flavonoid pro- anthocyanidins, quercetin, and kaempferol. Grape leaves are consumed in Middle Eastern cuisine and are an excellent source of vitamin K. Concord grape (Vitis labrusca) juice is high in vitamin C and polyphenols.

Grapefruit

Large yellow- or pink-fleshed citrus fruit that is an excellent source of vitamin C and a source of folate and potassium, whether its juice or the whole fruit is consumed. Naringin is a bitter principle of grapefruit and flavonoid that is partly responsible for grapefruit juice enhancing the bio- availability of certain medications.In laboratory studies, naringin exhibited antioxidant properties, inhibited tumor growth, and suppressed age-related blood pressure increases in hypertensive rats.

Green Leaf(y) Vegetables

Versatile family of plant foods, such as basil, arugula, and spinach, known for their antioxidant capacity. In particular, Brassica vegetables are sources of specific phytochemicals, such as flavonoids and glucosinolates. Generally, dark green leafy vegetables are noted for providing vitamin E, folate, magnesium, vitamin K, and chlorophyll. A high intake of green leafy vegetables was associated with a reduced risk of non-Hodg- kin's Lymphoma in a population-based case-control study (n=348 cases and 470 controls) in which a higher intake of green leafy vegetables and cruciferous vegetables was associated with a lower risk of non-Hodgkin's Lymphoma. Evidence was observed to support that higher vegetable consumption, including dark green vegetables and other plant foods, was significantly inversely associated with pancreatic adenocarcinoma.

Goji Berries

Goji berries are a nutrient-dense fruit rich in vitamins C, A, B2, and B3, along with antioxidants and fiber. Scientific research supports their potential benefits for antioxidant support, anti-inflammatory effects, eye health, and immune support. Including goji berries in your diet can contribute to overall health and well-being, especially when combined with a varied and nutritious diet.

Guava

Guavas are tropical fruits with edible seeds that are sources of polyphenols and are good sources of fiber (3 g, 10% DV) and vitamin A (343 IU, 11% DV), and excellent sources of vitamin C (125 mg, 138% DV) per 1 fruit. The 200 mg of vitamin C per 300 mL serving of guava juice was attributed with increasing the absorption of nonmeat-flesh iron, such as vegetable sources of iron. Preliminary research suggests guava lectin and galactose may have antidiarrheal properties.

Green Bean

Green beans are a nutritious vegetable rich in vitamins C, K, A, and folate, along with antioxidants and dietary fiber. Scientific findings support their potential health benefits, including antioxidant support, heart health promotion, and digestive health support. Including green beans in your diet can contribute to overall health and wellbeing, providing essential nutrients and supporting various bodily functions.

Gaseous state
Occurs when molecules are widely separated from each other as in air or water vapor.

Genetic code

The sequence of nucleic acid bases that code for proteins and give instructions to cells for how to grow and divide.

Glacial Acetic Acid

A highly concentrated solution of vinegar.

Glucose

A monosaccharide. The repeating unit in starches and cellulose. Also, together with fructose, glucose makes sucrose, table sugar.

Glycogen

Polysaccharide made by animals.

Gold

Element number 79 on the periodic table with the symbol Au. Gold has 79 protons, 79 electrons and 118 neutrons.

Guanine

One of the four nucleic acid bases that make DNA.

Gene

Short length of DNA that contains chemical instructions on how to build and run a living thing.

Genome

The entire set of genes and non coding DNA belonging to a living thing.

Hazelnut

Marble-sized nut, also called a filbert, that is a good source of protein, monounsaturated fatty acid, tocopherols, and phytosterols. In a laboratory study, hazelnuts 'phenolic compounds exerted antioxi- dant effects. In a 4-week single intervention study (n=21 normolipemic, healthy individuals), a hazelnut-enriched diet (1 g/kg/day) decreased the atherogenic tendency of LDL by reducing oxidized LDL levels and increasing vitamin E in LDL.

Hemicellulose

Type of dietary fiber that has characteristics of both insoluble and soluble fibers. It surrounds cellulose in plant cell walls. Legumes, fruits, and vegetables (particularly younger or less mature vegetables), and grains are the main sources of hemicellulose in the diet. Fibers in the large intestine promote stool passage and are associated with lower rates of diverticular disease, hemorrhoids, and appendicitis.

Honey

Intensely sweet viscous simple sugar, made of glucose and fructose, that is produced by bees from flower nectar. Flavors of honey vary depending upon the flower source. Use has been documented in the world's oldest medical literature for its topical healing properties due to its "high viscosity that helps to provide a protective barrier to infection." Honey contains the antibacterial protein royalisin. Four different varieties of honey exerted antimicrobial effects against S. Aureus in a laboratory study. Honey also appears to be at least as effective or more effective than the cough suppressant dextromethorphan in typical over-the-counter doses and the antihistamine diphenhydramine.

Honeycrisp Apple

Honeycrisp apples are a nutritious fruit rich in vitamins C, A, K, and E, along with antioxidants and dietary fiber. Scientific findings support their potential health benefits, including antioxidant support, digestive health promotion, and various positive impacts on overall health. Incorporating Honeycrisp apples into your diet can contribute to a balanced intake of essential nutrients and support overall well-being.

Honeydew melon

One-half cup of honeydew contains potassium (202 mg, 4% DV) and folate (17 µg, 4% DV), and is a good source of vita- min C (15.5 mg, 11% DV). Cucurbitacin-β is among its phytochemical constituents. Cucurbitacin-β has exhibited anti-hepatotoxic, anti-inflammatory, and anticancer properties in laboratory studies.

Hops

Flower of the hop plant; contains phytoestrogens and chalcones. The major bitter flavoring in beer; also used as a food preservative. Traditional uses of hops include as a sleep aid, stomachic, antibacterial, and antifungal agent7, sedative effects of hops have been observed in laboratory animals. In vivo and in vitro studies have demonstrated stomachic, antibacterial, antifungal, and cancer-preventative properties of hops.

Horseradish

Thick, woody root vegetable belonging to the Brassicaceae family that is a source of potassium and phytochemicals such as lutein, xeaxanthin, and glucosinolates. Its major flavoring constituent, allyl isothiocyanate, causes a burning sensation when it comes in contact with the mouth. Green-colored horseradish is frequently substituted for wasabi, the sushi condiment, in the US. Horseradish has been used to cure scurvy due to it vitamin C content (3.7 mg, 4% DV per 1 tablespoon). Horseradish exhibited antibacterial properties in an in vitro study. In a placebo-controlled clinical trial, horseradish also helped prevent urinary tract infections. The German Commission E has approved horseradish for oral use as "supportive therapy for infections of the urinary tract."

Huckleberries

Huckleberries provide vitamins C, K, B1, and B2, along with potent antioxidants and dietary fiber. Scientific findings support their potential health benefits, including antioxidant and anti-inflammatory effects, heart health promotion, and potential anticancer properties. Incorporating huckleberries into your diet, whether fresh or dried, can contribute to overall health and wellbeing, providing essential nutrients and protective antioxidants.

TOP 15 FOODS THAT KEEP YOU (H)YDRATED

1. Cucumber (96%)

If you love the crunch of a cucumber, you're in luck. Cucumbers have the highest water content of any solid food.

2. Iceberg Lettuce (96%)

Darker greens do provide more fiber, folate and vitamin K. But when it comes to staying hydrated, crispy iceberg has the goods.

3. Celery (95%)

In addition to being full of water, celery is a great source of fiber. Add some protein-rich nut butter and you have a healthy (and thirst-quenching) snack.

4. Radishes (95%)

This root veggie is full of flavor, vitamin C and fiber but low in calories. Add them to a green salad or grate them into a summer slaw.

5. Romaine Lettuce (95%)

This dark, leafy green packs in a lot of water but also provides many nutritional benefits. Romaine is a good source of vitamins C and A, folate and fiber.

6. Tomatoes (94%)

Many people think of tomatoes as vegetables, but they are actually the fruit with the highest water content. They also contain lycopene, which helps prevent cell damage.

7. Zucchini & Summer Squash (94%)
Whether you eat summer squash cooked or raw, it provides the same amount of water. So don't be afraid to throw some zucchini on the grill next to your turkey burger.

8. Asparagus (92%)
With its rough texture, most people choose to eat cooked asparagus. But its water content is the same whether you eat it cooked or raw, so go ahead and grill up some spears.

9. Bell peppers (92%)
All shades of bell peppers will quench your thirst, but green ones lead the pack in water content. As a bonus, bell peppers are high in antioxidants.

10. Cabbage (92%)
All common varieties of cabbage contain a lot of water raw and even more when cooked (94%). Some Chinese cabbages, such as bok choy, are 96% water and taste great tossed into a salad.

11. Cauliflower (92%)
If you love riced cauliflower, but don't like to eat it raw, you're in luck. Cooked cauliflower, like cabbage, provides even more water (93%) than eating it raw.

12. Mushrooms (92%)
To get the most water from your shrooms, eat them raw.

13. Spinach (92%)
Spinach is a great summer staple for salads and smoothies. In addition to its high water content, it's packed with nutrition including calcium, magnesium, iron and potassium.

14. Strawberries (92%)
These sweet berries are a great choice when you're hot and

sweaty. They're low in calories, but high in water, fiber and vitamin C.

15. Watermelon (92%)
It's no surprise that watermelon will hydrate you. But, like tomatoes and other red fruits and veggies, it's also a great source of lycopene.

Top humectant rich foods

Honey: Honey is a natural humectant and has been used for centuries in skincare for its moisturizing properties.

Glycerin: Though not a food in itself, glycerin is derived from fats and is used in many food products as a humectant. It's also commonly found in skincare products.

Aloe Vera: Aloe vera gel, extracted from the leaves of the aloe vera plant, is known for its hydrating and soothing properties.

Seaweed: Seaweed extracts are often used in skincare products for their moisturizing effects.

Oatmeal: When used in skincare, oatmeal can act as a humectant, helping to draw moisture to the skin.

Coconut Oil: Coconut oil is not only a good moisturizer but also has humectant properties, helping to keep moisture locked into the skin.

Cucumber: Cucumber has high water content and is often used in skincare to hydrate and soothe the skin.

Grapes: Grapes contain natural sugars and acids that can help attract and retain moisture when used in skincare.

Avocado: Avocado is rich in healthy fats and oils that can moisturize and hydrate the skin.

Jojoba Oil: Although technically a liquid wax, jojoba oil is often used in skincare for its moisturizing and humectant properties.

Vegan Histidine Rich Foods

Soybeans and Soy Products: Soybeans and foods made from them, such as tofu, tempeh, and soy milk, are good sources of histidine.

Legumes: Various legumes such as chickpeas (garbanzo beans), lentils, black beans, kidney beans, and pinto beans contain histidine.

Quinoa: Quinoa is a pseudo-cereal that is a complete protein source and contains histidine.

Pumpkin Seeds: Pumpkin seeds (pepitas) are a good source of histidine, along with other nutrients like zinc and magnesium.

Chia Seeds: Chia seeds are rich in protein and contain histidine, among other essential amino acids.

Sunflower Seeds: Sunflower seeds are another seed option that provides histidine, along with vitamin E and healthy fats.

Sesame Seeds: Sesame seeds are rich in histidine and are commonly used in various culinary preparations.

Oats: Oats contain some histidine and are also rich in fiber, making them a nutritious choice for breakfast or baking.

Buckwheat: Buckwheat is a gluten-free pseudo-cereal that contains histidine and other essential amino acids.

Spinach: While not as high in histidine as the above sources, spinach and other leafy greens contain small amounts of histidine along with other nutrients important for overall health.

Halogens
The group of elements in the column just left of the noble gases.

Helium
The second element on the periodic table. It has two protons, two neutrons, and two electrons. One of the noble gases.

Heterogeneous
A mixture of the "other kind" meaning that the mixture is not the same throughout.

Homogeneous mixture
A mixture of the "same kind" meaning that the mixture is the same throughout.

Hydrochloric acid
A strong acid found in the stomach of humans. For digestion. It is used to help break down food.

Hydrogen
The first element in the periodic table with the symbol H. It has only one proton, one electron, and no neutrons.

Hydroxide ion
One hydrogen atom and one oxygen atom bonded together.

HAZMAT suit
HAZMAT is short for hazardous materials. A HAZMAT suit keeps people safe from contagious microorganisms. It is worn by scientists and medical workers when working with dangerous pathogens.

Hot spring
Natural mineral water that bubbles up to the surface in volcanic regions.

Hydrothermal vent
A "chimney" on the seabed from which heated mineral water flows.

Hadron
Tiny particles of matter made of quarks bound together. Protons and neutrons are type of hadrons.

Iceberg Lettuce
Iceberg lettuce provides vitamins K, A, and folate, along with water and dietary fiber. While lower in vitamins compared to other greens, it contributes to hydration and digestive health. Incorporating iceberg lettuce into salads and sandwiches can

add crisp texture and contribute to overall dietary variety, but for higher nutrient intake, it is beneficial to also include darker leafy greens and a variety of vegetables in your diet.

Inulin

Fermentable carbohydrate (classified as a fructooligosaccharide or fructan) found in plant foods, such as asparagus, bananas, chicory, dandelion, garlic, Globe artichoke, Jerusalem artichoke, leeks, onions, wheat bran, and wheat flour, that is a prebiotic. Inulin improves laxation by increasing stool bulk, water content, and certain fecal bacteria, and may strengthen the intestinal epithelium possibly reducing the risk of gastrointestinal diseases. Gut microbiota improved in inulin supplemented (0.8 g/dL) formulafed newborn infants compared to breastfed infants in a double-blind, randomized, placebo-controlled clinical trial lasting for four months.

Iodine

The trace mineral referred to in the passage is iodine. Iodine is crucial for the synthesis of thyroid hormones T3 (triiodothyronine) and T4 (thyroxine), which play essential roles in regulating basal metabolism, growth, and body temperature. The passage discusses the historical significance of iodine in preventing goiter (thyroid enlargement) and the global issues related to iodine deficiency, as well as sources of iodine in the diet such as seafood, seaweed, iodized salt, and iodine fortified foods.

Iron

Crucial for hemoglobin in red blood cells and myoglobin in muscles, aiding oxygen transport for energy metabolism. Heme iron, found in animal flesh like meat, poultry, fish, and seafood, is well-absorbed (about 23%), while nonheme iron in eggs, legumes, and fortified cereals has lower absorption (2-20%). Iron deficiency, common in groups like toddlers, women, and adolescents, causes symptoms such as fatigue and decreased cognitive function due to insufficient intake or absorption issues.

Isoflavones

Plant compounds resembling estrogen that act on estrogen receptors in the body. They are found in foods like soybeans, tofu, and various fruits and vegetables. These compounds have mixed effects, potentially influencing cholesterol levels and menopausal symptoms like hot flushes. Research shows varying outcomes on their effectiveness, suggesting they may have benefits at higher doses for some symptoms, but overall results are inconclusive.

Isoprenoids

Phytochemicals such as menthol from peppermint oil, citral from lemongrass oil, and limonene from citrus rinds that impart flavors and fragrances. Fruits, vegetables, and cereal grains contain a variety of isoprenoid compounds. Isoprenoids possess anticancer activity in laboratory research.

Isothiocyanates

Phytochemicals formed from the breakdown of glucosinolates that are found in mustard seed and Brassica vegetables, such as watercress and broccoli sprouts. Isothiocyanates are compounds known for their antimicrobial properties against a broad range of pathogens. They also demonstrate anticancer activity in both cell cultures and animal models. Epidemiological studies exploring the link between phytochemicals and cancer risk have generally found mixed results. However, studies indicating an effect consistently show that higher dietary intake or levels of isothiocyanate metabolites in serum, plasma, or urine are associated with protective effects against lung cancer and gastrointestinal cancer.

Irish Moss

Irish moss contains small amounts of vitamins A and C and is rich in minerals like iodine, potassium, calcium, and magnesium. It is primarily valued for its culinary and industrial uses, particularly as a gelling agent and thickener in food products. While it has a historical use in traditional medicine, more scientific research is needed to fully understand its potential health benefits and nutritional value.

Isotope
Atoms of the same element that have the same number of protons but differing amounts of neutrons. Heavy isotopes are often radioactive.

Ionic bond
A chemical bond in which the electrons are not shared, but one atom donates electrons to the other atom.

Indigestion
Stomach pain caused by excess hydrochloric acid.

Immunity
The ability to resist infection by pathogenic microorganisms.

Infectious

Describes something that is capable of being passed from one person to another (such as a disease).

Infrared

Frequencies of electromagnetic radiation just below that of visible light

Jackfruit

Jackfruit provides vitamins C, A, and B6, along with antioxidants and dietary fiber. While not particularly high in vitamins compared to other fruits, jackfruit offers a range of potential health benefits and nutritional value. Including jackfruit in your diet, especially as a plantbased meat alternative, can provide essential nutrients and contribute to overall dietary diversity and health.

Jasmine Rice

Grown primarily in Thailand, 1 cup of white enriched jasmine rice is a good source of thiamin (0.31 mg, 25% DV), niacin (3.4 mg, 21% DV), iron (2.77 mg, 15% DV), and folic acid (180 µg, 45% DV), and supplies riboflavin (0.03 mg, 2% DV); and 1 cup of brown jasmine rice is an excellent source of iron (4.3 mg, 23% DV) and fiber (7.9 g, 28% DV). Rice is one of the longest cultivated cereal grains believed to have been grown for at least 5,000 years. Both white and brown rice are high glycemic index foods; people with diabetes may benefit from lower-glycemic-index foods, such as rice noodles. A questionnaire survey of 1,848 men and women between 20 and 60 years of age found that rice, possibly due to its high glycemic index, was associated with better sleep quality, while bread and noodles were associated with poor sleep quality.

Jalapeno

Jalapeños provide vitamins C, A, K, and B6, along with antioxidants and capsaicin. Scientific findings suggest potential health benefits, including antioxidant support, digestive health promotion, and potential cancer-fighting properties. Incorporating jalapeños into your diet, if tolerated, can add flavor and nutritional value to meals while potentially supporting overall health and wellbeing.

Jerusalem Artichoke

Similar to water chestnuts in taste, Jerusalem artichoke carbohydrate is primarily inulin, an indigestible carbohydrate that is fermented in the colon, versus starch that is digested and absorbed. Fructans from Jerusalem artichokes reduced insulin response, compared to fructose, an effect attributed to fructans' slowing of GI transit time, in a small (n = 8) trial of healthy subjects.

Jicama

Jicama contains fewer than 25 calories per 1/2 cup, 2.9 g fiber (10% DV), including lignin (a type of fiber), and phenolic compounds.In animal studies, jicama reduced serum glucose and increased insulin sensitivity.

Juniper Berry

It contains phenolics and other phytochemicals, including limonene and myrcene. Juniper berry provides the characteristic flavor of gin. Juniperus communis has been used as a traditional cure for tuberculosis and other respiratory diseases. Juniper has in vitro antioxidant properties that inhibit lipid peroxidation. Juniper berry was hepatoprotective in an animal model.

The Top Foods for Healthy Joints

Seeds and Nuts

Seeds and nuts are packed with healthy Omega-3 fatty acids known to fight inflammation and help reduce it in your connective tissue and joints. Great options include almonds, walnuts, pine nuts, flax seeds, and chia seeds.

Fruit

Many fruits have powerful antioxidants that reduce inflammation in the body which helps joint pain. Blueberries are one that have strong flavonoids that turn the inflammatory response in your body off. Pineapple also contains a strong element, bromelain, which is shown to relieve joint pain that comes with rheumatoid arthritis and osteoarthritis. Tomatoes have the antioxidant lycopene that provides improvements for this physical health concern, too.

Cruciferous Veggies

Cruciferous veggies include brussels sprouts, cauliflower, and broccoli. These foods have been found to block enzymes that result in swelling of the joints. They're also packed with a great dose of vitamins and minerals.

Beans and Lentils

Pinto beans, chickpeas, black beans, soybeans, and lentils

all include anthocyanins, a flavonoid that helps reduce inflammation in your body. Beans and lentils also provide a great source of essential minerals, fiber, and protein.

Olive Oil

Oils such as peanut oil, vegetable oil, and sunflower oil can increase inflammation levels. Olive oil, however, is an excellent substitute for salad dressings or cooking. It's a healthy fat and packed with those inflammation fighting Omega-3s.

Whole Grains

While the proteins in refined grains can trigger the body's inflammatory response, whole grains may help counteract it. Grains recommended for reduced inflammation and joint pain includes whole oats, rye, barley, and whole wheat.

Root Veggies and Garlic

The aromatic root vegetables such as onions, garlic, turmeric, and ginger are all known for their anti-inflammatory properties. They can treat joint pain and other symptoms of arthritis. Root veggies and garlic can be added to meals for additional flavor while boosting joint health.

Dark Chocolate

Dark chocolate is delicious and great for joint paint because cocoa contains antioxidants that counteract inflammation. The key is to choose chocolate with a high percentage of cocoa and to indulge in moderation.

Kaempferol

Common dietary flavonoid found in a wide variety of plant

foods such as tea, broccoli, cabbage, kale, beans, endive, leek, tomato, and strawberry. Consuming foods containing kaempferol was associated with a reduced risk of developing cancer and cardiovascular disease in epidemiological studies. Kaempferol was one of several flavonoids associated inversely with lung cancer among tobacco smokers, but not among nonsmokers in a population based case control study (n=558 lung cancer cases and a group of 837 controls).

Kale

Kale is a dark green leafy Brassica vegetable known for its nutritional richness. Per 1/2 cup serving, kale provides calcium (45 mg, 3% DV), folate (10 µg, 2.5% DV), magnesium (11 mg, 2% DV), and is exceptionally high in vitamins: vitamin A (258 µg RAE, 29% DV), vitamin C (40 mg, 44% DV), and vitamin K (273 IU, 227% DV). Kale also contains phytochemicals such as lutein, zeaxanthin, and allyl isothiocyanate. Kale seeds exhibited antioxidant properties in vitro.

Kefir

Kefir provides vitamins B12, K2, D, biotin, and folate, along with probiotics that support gut health and overall wellbeing. Scientific findings support its role in promoting digestion, enhancing nutrient absorption, and potentially reducing inflammation. Including kefir as part of a balanced diet can provide essential vitamins, probiotics, and potential health benefits, especially for those looking to support gut health and immune function.

Key Lime

A good source of vitamin C and flavonoids. Lime essential oil from lime peel reduced body weight in laboratory animals. Key lime inhibited the growth of antibiotic resistant H. pylori in experimental research.

Kiwano

Kiwano provides vitamins C, A, E, and B6, along with antioxidants and hydration due to its high water content. While specific scientific findings on kiwano are limited, its nutritional profile suggests potential health benefits similar to other vitamin-rich fruits. Including kiwano in your diet can add variety and provide essential nutrients, supporting overall health and well-being.

Kiwifruit

Also called kiwi. Small green or goldfleshed fruit that is eaten fresh and depending on the variety, is a source of vitamins C and E, phenolics, carotenoids, lutein, and xeaxanthin. When its small, edible black seeds are consumed, one small, whole kiwifruit supplies 2 g fiber. In experimental research, kiwifruit improved markers of immune function in mice. Kiwifruit polysaccharides increased fibroblast activity in an in vitro study, which may have implications for human collagen synthesis. Kiwi may improve laxation, aid digestion, and promote healthy gut microflora in humans, and, in addition, has antioxidant and antiplatelet-aggregatory properties.

Kombucha

It has resurged to popularity in recent years due to its purported functional properties including anti-inflammatory and radical scavenging. Limited evidence suggests kombucha tea may offer benefits similar to probiotic supplements, including promoting a healthy immune system and preventing constipation. At present, however, valid medical studies of kombucha tea's role in human health are very limited and its benefits or side effects have not been widely reported.

Kumquat

Citrus peel is used in traditional medicine for digestion, severe cold, and fever. Kumquats are a source of flavones, vitamin C, and the carotenoids alpha-carotene, beta- cryptoxanthin, lutein, and xeaxanthin. Kumquats can be eaten raw and whole or can be used to make marmalade. Canned, peeled kumquats are often served as desserts in Chinese restaurants. Kumquat peel and pulp contain phenolic compounds and flavonoids.

Kinesin

A protein molecule that is used to carry a cargo down a microtubule "road" to various parts of the cell.

Lactose

The disaccharide in question is lactose, exclusively found in milk and present in smaller amounts in yogurt and cheese. It's widely used as a cheap bulking agent in processed foods, often hidden under names like sodium caseinate. Lactose intolerance, affecting up to 75% of people globally, occurs when the enzyme lactase is insufficient to break down lactose in the small intestine. This leads to undigested lactose fermenting in the colon, causing symptoms like abdominal pain and diarrhea. Despite these issues, lactose aids calcium and magnesium absorption and acts as a prebiotic, supporting beneficial gut bacteria growth. Avoiding lactose-containing foods may reduce calcium and vitamin D intake, impacting bone health, particularly in lactose-intolerant individuals.

Legume (Pulse)

The vegetable in question is a type of legume, such as lentils, beans, or chickpeas. These are packed with nutrients like fiber (7 g), iron (5 mg), magnesium (109 mg), protein (7 g), and folate

(60 µg) per halfcup serving of baked beans. They also provide zinc, copper, and manganese in smaller amounts.

Legumes contain various beneficial compounds like phytochemicals and are a good alternative protein source to meat, fish, or tofu according to USDA guidelines.

Research suggests that regular legume consumption may lower overall mortality and reduce cancer risk. They also show promise in managing blood sugar levels, particularly beneficial for individuals with diabetes due to their high soluble fiber and low glycemic index.

Despite these benefits, legumes are not commonly consumed daily by most Americans, highlighting a missed opportunity for improving nutrition and health.

Lemon

Its citrate content is the highest among citrus fruits.16 Lemon juice is a good source of vitamin C (11.8 mg, 13% DV per 1 tablespoon) and supplies liminoid, beta-cryptoxanthin, lutein, xeaxanthin, coumarins, hesperetin, quercetin, and myricetin.

Lemon essential oil, derived from lemon peel, has been effective in reducing airborne bacteria when used as an aerosol. Studies on lemonade consumption have shown mixed results: while it increased urine volume in individuals prone to kidney stones, it did not consistently improve urinary citrate or uric acid levels across trials. Longterm studies indicate that lemonade therapy can significantly reduce kidney stone formation in certain patients over extended periods.

Lemon peel contains coumarins that inhibit free radical formation, offering potential antioxidant benefits. Hesperidin, found in lemon peel, has demonstrated anti-inflammatory and cardiovascular protective effects in both experimental and clinical settings. It shows promise in treating conditions such as hemorrhoids and venous stasis by reducing capillary leakage.

Lemongrass

Lemongrass tea has been used in traditional medicine to treat hypertension and diabetes. Lemongrass essential oil, when sprayed as an aerosol, reduced airborne microorganism contaminants. Laboratory research suggests that anti-inflammatory compounds in lemongrass work by inhibiting the release of pro-inflammatory cytokines. Lemongrass attenuated animal liver dam- age in an experimental study.

Lentil

Lentils provide vitamins B9, B1, B3, and B6, along with dietary fiber and antioxidants. Scientific findings highlight their nutritional value and potential health benefits, including support for heart health, blood sugar regulation, and digestive health. Including lentils in your diet can contribute to meeting daily nutrient needs and promoting overall well-being, especially as a sustainable and versatile plant-based protein source.

Lepton
A small fundamental particle affected by weak forces. Electrons and neutrinos are leptons.

Lettuce

Salad green, such as the Bibb variety, that grows in small, loose heads and is an excellent source of vitamin K (56 μg, 46% DV per 1 cup); lola rosa (also spelled lola rossa and lolla rosa), a mild-flavored, looseleaf Italian variety with ruffled burgundy leaves that is frequently included in spring mix and is a rich source of anthocyanins that are concentrated in its intense red-pigmented parts; and lollo biondo, the green-leafed sister of lola rosa, that has been described as "pleasantly strong and nutty," although slightly bitter. Lettuce intake correlated with higher serum carotenoid concentrations, specifically of xeaxanthin and beta-cryptoxanthin, in postmenopausal osteoporotic women (n=59), a subgroup that may have compromised antioxidant defenses.

Lotus Root

Lotus root provides vitamins C, B6, B1, and B2, along with dietary fiber and antioxidants. While not particularly high in vitamins compared to other vegetables, lotus root offers nutritional benefits and potential health advantages, including digestive support and antioxidant protection. Incorporating lotus root into your diet can add variety, texture, and essential nutrients to meals, contributing to overall health and well-being.

Lingonberry

Phytoestrogen that is abundant in the Western diet as are its metabolites enterolactone and enterodiol. Lignan is found in flaxseed, pumpkin seed, sesame seed, soybean, broccoli, whole grains, beans, peas, and some berries. A meta-analysis of 21 epidemiologic studies concluded that high lignan exposure is associated with a reduced breast cancer risk in postmenopausal women. GI microorganisms interacting with lignan enterodiol and enterolactone have generated bioactive compounds that retarded experimentally induced cancer. "Ecological data suggest a long-term diet high in plant material rich in biologically active compounds, such as the lignans, can significantly influence the development of prostate cancer over the lifetime of an individual."

Lipid

Found in plant oils and animal fats, its roles in foods include improving palatability, flavor, and aroma; serving as a source of essential fatty acids and energy and as a source of fat soluble vitamins and phytochemicals.

Lutein

A xanthophyll carotenoid found abundantly in green vegetables alongside zeaxanthin, gives these foods their yellow color and plays a crucial role in protecting the macula of the retina. A higher intake of lutein and zeaxanthin is associated with lower risks of age-related macular degeneration (AMD), a leading cause of blindness in older adults.

While studies show mixed results on whether dietary lutein and zeaxanthin reduce early AMD risk, they consistently link these carotenoids to a lower risk of advanced AMD. However, supplements containing lutein have shown limited effectiveness in preventing AMD progression in randomized trials.

Consuming foods rich in lutein and zeaxanthin, around 6.9 to 11.69 mg daily, as estimated from dietary surveys, may modestly reduce the risk of cataracts. Vegetables like broccoli, spinach, and kale are associated with lower cataract risk based on large studies involving both men and women.

Lychee Fruit

Raw lychee fruit contains vitamin C (6.9 mg, 7% DV) per 1 fruit, and is a source of flavonoids, sterols, and triterpenes. Lychee has been used in traditional Asian medicine for the treatment of a cough, flatulence, stomach ulcers, diabetes, obesity, testicular swelling, hernia like conditions, epigastric and neuralgic pains; and it has been used as a hypoglycemic, cancer-preventative, antibacterial, antihyperlipidemic, antiplatelet aggregatory, antitussive, analgesic, antipyretic, hemostatic, diuretic, and antiviral agent. Polyphenolic compounds from lychee fruits were strong antioxidants in laboratory studies. Lychee flavonoids exerted immunomodulatory and anticancer activities in vitro.

Lycopene

Carotenoid antioxidant that imparts pink and red pigment to certain fruits and vegetables such as watermelon, pink grapefruit, guava, and tomatoes. Approximately 85% of the dietary lycopene intake in the US is from tomatoes (Table L.3). Lycopene, found in tomatoes, shows promise in reducing cellular oxidative stress in lab studies. Observational research suggests higher blood levels of lycopene might lower cancer risk, but FDA reviews found no clear link to reduced risk for various cancers.

In prostate cells, lycopene appears to have anti-proliferative, antioxidant, anti-inflammatory, and hormone-regulating effects. Meta-analyses indicate a potential inverse relationship between lycopene intake and prostate cancer risk, especially with frequent tomato-based food consumption.

However, large trials like the Prostate Cancer Prevention Trial found no link between dietary lycopene or supplements and prostate cancer risk. Similarly, trials on PSA levels showed no significant benefits from lycopene supplements. Overall, more robust evidence is needed to support using lycopene for preventing prostate cancer or affecting PSA levels.

Top Foods That Increase Libido (Black Raspberries - Saffron):

Black Raspberries

Both the berries and the seeds will transform your mindset when you're trying to get in the mood. "This phytochemical-rich food enhances both libido and sexual endurance," say Anna Maria Gahns-Clement, Ph.D., and Brian Clement, Ph.D., the authors of 7 Keys to Lifelong Sexual Vitality and directors of Hippocrates Wellness in West Palm Beach, Florida. Consume some black raspberries or a tablespoon of seeds a few hours before you plan to get busy.

Broccoli

Raw, sautéed, or cooked, broccoli in your salad or alongside your steak can help to increase your libido. "I suggest it here because of the high vitamin C content," says Keri Glassman, R.D., C.D.N., a registered dietitian and founder of Nutritious Life. Broccoli is one of the best foods to make you sweet in bed and increase your libido. "Vitamin C aids in blood circulation to organs and has also been associated with an improved female libido," she adds.

Cloves

This sex superfood is versatile when it comes to cooking: It can be brewed in hot apple cider, infused into your favorite dish, or added to a chai tea latte. Just make sure to share it with your partner, too. "In India, cloves have been used to treat male sexual dysfunction for centuries," says Glassman. Research published in the journal BMC Complementary and Alternative Medicine agrees, discovering that clove extracts produced an increase in the sexual activity of normal male rats.

"Cloves are also used to rid bad breath, which can't hurt your kissing skills either," says Glassman. Try tossing in powdered cloves in Mexican food next time, too, she suggests. Add a little cumin and cinnamon and you've got a tasty, multifaceted aphrodisiac.

Figs

Want to be completely irresistible the next time your partner sees you? There's a simple food solution: Figs. "They're considered excellent stimulants of fertility and enhance the secretion of pheromones," say Gahns-Clement and Clement. Feast on the fruit before getting it on and find out if you're a little sweeter in bed.

Watermelon

Chocolate has the reputation of being the aphrodisiac of choice for those with a sweet tooth, but researchers at the University of Guelph in Canada found that this is most likely just a myth. Though ingredients such as phenylethylamine in chocolate may boost serotonin and endorphin levels, there is no link between chocolate and improved sexual performance or arousal. So if you're really looking for a sweet libido-booster, stick to a slice of watermelon.

Although it's 92 percent water, that remaining 8 percent of fruit is jam-packed with vital nutrients for sexual health. Watermelon "has ingredients that deliver Viagra-like effects to the body's blood vessels and may even increase libido," according to research from the Texas A&M Fruit and Vegetable Improvement Center. More specifically, "watermelon contains a phytonutrient called citrulline, which the body converts to arginine, an amino acid that boosts nitric oxide levels in the body, which relax blood vessels in the same way a medicine like Viagra does," say Gahns-Clement and Clement.

Eggs

Poached, scrambled, fried, deviled. No matter how you make them, a couple of eggs will rev you up after a long day so you can go the distance in the dark. "Eggs are high in protein, which is a source of stamina, and they're also low in calories," says Glassman. In addition, they're a good source for the amino acid L-arginine, which has been shown effective in treating types of heart ailments and erectile dysfunction.

Ginseng

Ginseng is one of the scientifically proven foods that increase libido in women: Researchers at the University of Hawaii found that women who took a ginseng-rich supplement showed a significant increase in libido in a month, and 68 percent also said their overall sex life improved dramatically.

"Add ginseng into your diet or try one of the many ginseng teas available. Just don't jump at the sight of ginseng, though. Many energy drinks that claim to have ginseng in them also contain chemicals and tons of sugar," advises Glassman.

Lettuce

A small salad with oil and vinegar as dressing might amp up your sex drive. "Iceberg lettuce contains an opiate that helps to activate sex hormones," say Gahns-Clement and Clement. Nosh on a bowl at dinner and you'll be ready to go by the time you turn in for the night.

Ginger

"King Henry VII and the ancient [Asian cultures] were astute when using ginger for medicinal purposes: In the 21st Century, those of us who know about botanical-ceuticals know that ginger helps circulation, temperature adjustment, and mucoid detoxification (mucus-like residue that can coat your GI tract), and is also a libido enhancer," note Gahns-Clement and Clement. Whether raw, in supplement form, or added to your favorite recipe or drink, ginger also lends itself to defense against winter's hard cold and flu season, which is good because trying to get frisky with a runny nose never ends well.

Saffron

If you want to splurge on foods that make women sweet in bed and increase libido, look no further than saffron. Though on the pricey side, saffron is a treat for your stomach and sex life: Researchers at the University of Guelph found that saffron can improve sexual performance. To use, soak the threads in hot liquid for 15 minutes then add to any grain such as rice, quinoa, or barley or use it in a soup or stew, suggests Glassman.

Lead

Element number 82 on the periodic table with the symbol Pb. Lead has 82 protons, 82 electrons and 125 neutrons.

Liquid state

The fluid state of a substance.

Lepton

A member of a family of fundamental particles that includes electrons and neutrinos. Leptons are affected by weak force.

Magnesium

Magnesium (Mg) is a vital mineral essential for energy metabolism, vitamin D action, protein synthesis, and many other bodily functions. It regulates blood glucose, blood pressure, and plays a key role in nerve function, muscle contraction, and heart rhythm.

Most magnesium is stored in bones, with the rest in soft tissues.

It's often under-consumed, especially by older men and teenage girls. Good sources include vegetables, dairy, whole grains, and proteins, as well as mineral rich water.

Deficiency can result from poor intake, health conditions like diabetes, or medication use. Symptoms range from muscle cramps to more serious issues like abnormal heart rhythms.

Ensuring adequate magnesium intake, around 500-1000 mg daily, may help lower blood pressure, reduce diabetes risk, and support overall health, including bone strength and metabolic balance.

Mango

Juicy, orange-fleshed tropical fruit eaten fresh or dried, that is an excellent source of vitamin A (893 IU, 29% DV) and supplies polyphenols, terpenoids, steroids, phenolics, and flavan-3-ols. In in vitro and in vivo models, mango has exerted antioxidant, iron chelator, antiinflammatory, antinociceptive, antitumor, and immunomodulatory properties. Raw M. indica fruit killed aerobic and anaerobic acne inducing bacteria and was antiinflammatory in skin cells.

Mediterranean Diet

Traditional dietary style of the Mediterranean basin that is associated with cardiovascular and cognitive health. It includes large amounts of olive oil, vegetables and leafy greens, fruit, whole grains, nuts, legumes, moderate amounts of fish, meat, dairy products, and red wine, and is low in eggs and sweets. It is high in unsaturated fat and polyphenolic compounds, such as flavonoids and phytosterols from red wine, olive oil, coffee, tea, nuts, fruits, vegetables, herbs, and spices. A metaanalysis of 11 randomized controlled trials (RCTs) of healthy adults at high risk of cardiovascular disease (CVD) (n = 52,044) found limited but favorable evidence of Mediterranean dietary patterns on CVD risk factors. A meta-analysis (n => 4.7 million people) found reduced mortality, CVD-related mortality, and a reduced risk of Parkinson's and Alzheimer's disease related to fol- lowing a Mediterranean Diet-style eating pattern (Table M.2).

Melatonin

The naturally produced melatonin hormone regulates circadian rhythm. Older adults may be prone to develop disorders related to an altered circadian rhythm since levels decline with age. Studies conducted to evaluate the efficacy of melatonin and sleep used chemosynthesized melatonin or melatonin sourced from the pineal gland of beef cattle. In laboratory studies, melatonin suppressed tumor angiogenesis and scavenged hydroxyl radicals. Melatonin supplementation was not effective in treating most primary sleep disorders with short-term use, a comprehensive review of randomized clinical trials found. A dose of 0.1 mg to 50 mg/kg orally was the dosage used in 16 RCTs to mimic normal physiological circadian rhythm.

Miso

Miso provides vitamins K, B2, B3, and folate, along with probiotics and potential antioxidant benefits. Scientific findings support its role in promoting gut health, enhancing nutrient absorption, and potentially contributing to overall well-being. Incorporating miso into your diet as a seasoning or in soups and sauces can add flavor, nutrients, and potential health benefits, especially for digestive and immune health.

Monounsaturated Fatty Acid

Dietary triglyceride whose fatty acids predominately contain one double bond. Monounsaturated fatty acid is the major triglyceride found in canola, olive, peanut, and sunflower oils; avocados; and nuts and seeds. The Mediterranean Diet is high in monounsaturated fat. Replacing saturated and trans fats with monounsaturated and polyunsaturated fats reduces LDL cholesterol.

Molasses

A thick, dark syrup produced during the refining of sugar cane or sugar beets. It is rich in minerals like iron, calcium, and potassium, and is used as a sweetener in baking and cooking, as well as in marinades and sauces.

Mulberry

An excellent source of vitamin C (25 mg, 27% DV) and potassium (750 mg, 22% DV) per 1/2 cup, mulberries also contain 5 g (17% DV) of fiber, and black or red mulberries are a source of anthocyanin and resveratrol. "Morus alba L. has long been used in traditional oriental medicine." In laboratory studies, M. nigra exhibited neuroprotective and membrane-protective, antioxidant effects; resveratrol ameliorated stress-induced Irritable Bowel Syndrome-like behaviors by regulating the brain-gut axis, and was anti-inflammatory; additionally, resveratrol, the phytochemical attributed with the cardioprotective effects of red wine in the French diet, modulated synthesis and secretion of lipoproteins, and arrested tumor growth as well as inhibited carcinogenesis in different experimental models; and in healthy humans promoted nitric oxide production by vascular endothelium and inhibited the synthesis of thromboxane in platelets and leukotriene in neutrophils, hindering platelet aggregation.

Mung bean

In addition to being excellent sources of fiber (8 g, 28% DV) and protein (12 g, 24% DV) per 1/4 cup, mung beans are a source of phenolics, flavonoids, and terpenoids. Per 1/2 cup, mung bean sprouts supply 1 g fiber (5% DV) and 2 g protein (4% DV), therefore are not appreciable sources of these important macronutrients. Mung bean sprouts and seed extracts exerted antidiabetic effects in animals with Type II diabetes, and antioxidant effects in laboratory research. Extracts made from mung bean sprout exerted antitumor effects in laboratory studies. Mung beans possess antioxidant activity.

Mushroom, maitake

Maitake mushroom contains 22 calories per 1 cup and supplies 4 mg of niacin (20% DV), 786 IU of vitamin D (200% DV), and small amounts of folate (21 μg of folate, 5% DV) and carbohydrate (5 g) including the polysaccharides lentinan and D-fraction. Maitake mushroom polysaccharides exerted antitumor and immune activity in laboratory studies. Maitake mushroom exhibited antidiabetic properties in labo- ratory analyses and in two cases of patients with Type II diabetes.

Mushroom, oyster

Oyster mushrooms exhibited anti-inflammatory properties in laboratory studies. Antitumor and immunomodulating factors have been isolated in oyster mushrooms.

Mushroom, shiitake

Also spelled shitake. Brown and white fungus that has a meat-like texture and is a source of riboflavin (0.082 mg, 6% DV), niacin (1.47 mg, 9% DV), vitamin B6 (0.11 mg 6% DV), and vitamin D (7 IU, 2% DV). Chemical analysis of shiitake mushrooms have isolated antibiotic, anticarcinogenic, and antiviral compounds. In in vitro and animal studies, shiitake constituents, such as lentinan, a polysaccharide flavonoid, have demonstrated immunomodulatory, antihypertensive, and antitumor effects, Shiitake mushroom alkaloids have exhibited anticarcinogenic properties in laboratory studies.

Mustard seed

In a laboratory study, mustard exhibited chemoprotective properties, which were attributed to its allyl isothiocyanate content. Flavonoids have been isolated in shoots, roots, and extracts of Brassica alba. Mustard seed suppressed psoriasis like inflammation in laboratory animals.

Matter
A general term for the stuff that makes up both living and nonliving things.

Mendeleev Dmitri Ivanovich
The scientist to organize the elements into an arrangement that would eventually become the Periodic Table of Elements.

Micelle
The small droplets of oil and soap in a water solution.

Microtubule
A molecule road inside cells that kinesis uses to move cargo.

Mineral oil
Octadecane.

Mixture

Any substance that is made of more than one thing.

Molecule(s)
The product of two or more atoms put together with bonds.

Monosaccharide
Any of a number of compounds that is a single sugar.

Meson
A subatomic particle made of two quarks.

Microbe
A non scientific term for a microscopic living thing, often used to mean germ.

Molecule
A substance made from two or more atoms bonded together and that gets involved in chemical reactions.

Monocyte
The largest white blood cell and part of the body's immune system.

Magnetic field
Produced by moving charged particles, changing electric fields, or the spinning electrons of a magnetic material. A magnetic field generates a force felt by charged objects and other magnets.

Mechanical advantage
The measure of how much a tool or machine boosts force.

Mass
A property of matter that describes how strongly it is affected by the force of gravity.

Matter
The fabric of the stuff that makes up the world around us; the things that we can touch, smell, and see. Matter is made up of atoms, which in turn are made up of smaller, elementary subatomic particles.

Microsecond

One millionth of a second.

Momentum

A measure of the oomph that a traveling body, such as a particle, has. A particle's momentum s its mass multiplied by its speed.

Natto

A traditional Japanese food made from fermented soybeans. It has a strong flavor and distinctive slimy texture due to the production of sticky threads of beneficial bacteria during fermentation. Natto is often eaten with rice and other toppings as a breakfast dish in Japan, known for its probiotic properties and nutritional benefits.

Nectarine

Both the white and orange cultivars supply fiber (2.4 g, 8.5%), vitamin C (7.7 mg, 27.5%), potassium (285 mg, 6% DV) and cyanidin, a proanthocyanidin compound. The orange cultivar is a good source of vitamin A (471 IU, 15% DV) and carotenoids such as beta-carotene and beta-cryptoxanthin. Epidemiology suggests that diets high in carotenoid-rich fruits reduce the risk of oxidation-dependent diseases, such as cancer, atherosclerosis, and macular degeneration. In laboratory studies, proanthocyanidins demonstrated their free radical scavenging, antibacterial, antiviral, anti- carcinogenic, anti-inflammatory, antiallergic, and vasodilatory actions; in addition, they inhibited lipid peroxidation, platelet aggregation, and capillary permeability and fragility, and demonstrated their antithrombotic potential by inhibiting cyclooxygenase and lipoxygenase of platelets in vivo.

Niacin

Niacin, also known as vitamin B3, is a water-soluble vitamin crucial for energy metabolism. It can be found in protein-rich foods like legumes and meat, grains (both enriched and whole), and vegetables such as mushrooms, potatoes, and asparagus.

Historically, niacin deficiency, known as pellagra, was prevalent in the Southern US after the Civil War due to a diet heavily reliant on niacin-deficient corn. Pellagra symptoms include dermatitis, diarrhea, dementia, and neurological issues.

Inadequate intake of iron, riboflavin, or vitamin B6 can hinder the conversion of tryptophan (an amino acid) to niacin. Alcoholism can also lead to pellagra by causing malnutrition and inhibiting this conversion process.

High doses of niacin are known to lower LDL cholesterol and triglycerides while raising HDL cholesterol significantly.

Nitrates and Nitrites

Nitrites are naturally occurring compounds linked to the nitrogen cycle in soil and water, present in low concentrations in soil, water, all plants, and meats. Cured meats are a major dietary source, where sodium nitrite is added for preservation and flavor. Nitrates, found in fruits and vegetables, are also present in municipal and well water as a pollutant. Vegetables are the primary source of dietary nitrate.

When ingested, nitrates are converted to nitrites, which can bind to hemoglobin and form methemoglobin, affecting blood oxygen levels at high concentrations. Nitrites and nitrates, along with their derivatives called nitrosamines, are associated with gastrointestinal cancers. Despite these risks, they also have beneficial effects on vascular and immune systems.

Regulatory agencies like the EPA and FDA oversee nitrate and nitrite levels in water and certain foods. Dietary intake of nitrites and nitrates from vegetables and fruits contributes to the blood pressure-lowering effects of diets like the DASH diet. In animal studies, nitrites have shown to enhance mucosal blood flow, act antimicrobially, protect against heart attacks and strokes, and reduce vascular inflammation from highfat diets. Certain probiotic bacteria can neutralize the effects of nitrosamines.

Noni

There are anecdotal reports of the successful use of noni to treat colds and flu, but the FDA has warned several noni product manufacturers to stop making false claims related to noni product use. Noni was shown in laboratory studies to have antioxidant, immune stimulating, tumor-fighting, anti-inflammatory, antioxidant, smooth muscle stimulatory, and histaminergic effects.

Nori

Seaweed algae that is a source of calcium, iron, iodine, magnesium, zinc, potassium, sodium, and the phytochemicals chlorophyll, beta-carotene, and lutein. Raw nori contains vitamin B12; however, dried nori is not a reliable source of vitamin B12. In a laboratory study, nori was antimutagenic.

Nopal (Prickly Pear Cactus)

Commonly used in Mexican cuisine, nopal refers to the pads or stems of the prickly pear cactus. They are often grilled, boiled, or used fresh in salads and salsas, known for their slightly tart flavor and mucilaginous texture. Nopal is also valued for its nutritional benefits, including high fiber content and various vitamins and minerals.

Nut

Protein-rich (7 g of protein, 14% DV per tablespoon or ½ oz) legume (e.g., the peanut) or drupe (e.g., almond or walnut) that is low in saturated fat, high in fat (containing 44% to 76% total fat which is mostly unsatu- rated) and calories (85 calories per 1/2 oz), and contains fiber (3 g, 10% DV per 1/2 oz). In a large prospective cohort study (n = 86,016 women in The Nurses 'Health Study) women who ate more than five oz of nuts per week had a significantly lower risk of coronary heart disease (CHD) than those who rarely or never consumed nuts. Five oz of any type of nut per week reduced risk of CHD in women. Consumption of 50–113 g (1/2 to 1 cups) of nuts daily with a diet low in saturated fat and cholesterol decreased total cholesterol by 4–21% and LDL cholesterol by 6–29% when weight was not gained.

Nutmeg

Nutmeg has been used as an antiflatulent in traditional medicine. Consuming intact nutmeg nuts has been known to cause lethargy, intoxication, and hallucinogenic effects dating back to early medical literature. In laboratory studies, nutmeg improved insulin sensitivity and secretion, and inhibited intestinal alpha-glucosidase leading to a slower postprandial glucose response. Myristicin exhibited hepatoprotective effects in experimental research. Elemicin was antimicrobial against the human enteropathogen Campylobacter jejuni in a laboratory study.

Neutralization reaction another name for an acid-base reaction.

Neutralize

The result following an acid base reaction; a neutral solution is neither acidic nor basic.

Neutrons
One of the three fundamental particles that make atoms. It has an atomic mass of 1 and carries no charge it is neutral.

Nitrogen
The seventh element in the periodic table. Nitrogen has the symbol N with 7 protons, 7 electrons and 7 neutrons.

Noble gases
The group of elements on the far right hand side of the periodic table. These elements do not react easily with any other element.

Nucleus
The central portion of an atom that houses the protons and neutrons.

Nuclear
Anything relating to the nucleus of an atom. Nuclear fusion is the joining together of atomic nuclei, and nuclear fission is the process of breaking a nucleus apart.

Neutral
Having no charge. A neutral particle does not interact with, and is not affected by, electromagnetic force.

Nucleus
The central part of an atom, made up of protons and in all elements apart from hydrogen—neutrons.

Oat

Versatile whole grain known for its content of betaglucan, a soluble fiber found in oat endosperm cell walls. Oat betaglucan is also called oat bran soluble fiber, oat fiber, oat soluble fiber, and oat bran fiber, though betaglucan is obtainable through other foods, such as barley. Oatmeal and whole oats provide all B vitamins except B12, and in addition, iron and phytochemicals called avenanthramides. In the National Health and Nutrition

Examination Survey 2001–2010 (n = 22,823), oatmeal consumers had higher dietary quality and lower body weights, waist circumferences, and body mass indices than oatmeal non-consumers. Whole and processed oats can significantly reduce postprandial blood glucose.

Okra

The seeds and skin contain flavonols, such as quercetin, flavonol derivatives, such as catechin, and polyphenolic compounds, such as hydroxycinnamic acid. A. esculentus reduced blood glucose and blood lipids in diabetic rats.

Olive

Greek kalamata olives (Olea euro- paea L. cv. Kalamata) contain the polyphenolic compound oleuropein, responsible for the bitterness in olives, and its metabolite, hydroxytyrosol, in addition to being an excellent source of monounsaturated fatty acids (MUFAs). As shown in laboratory studies, kalamata olives contain antioxidant, antiinflammatory, hormone-like, anticarcinogenic, antimicrobial, antihypertensive, antidyslipidemic, laxative and antiplatelet constituent. Hydroxytyrosol was neuroprotective in an animal model of dementia, and "olive oil phenols have shown neuroprotective effects against cerebral ischemia, spinal cord injury, Huntington's disease, Alzheimer's diseases, multiple sclerosis, Parkinson's disease, aging, and peripheral neuropathy in vitro and in vivo.

Olive Oil

High-MUFA oil expressed from olives is a major fatty component of the Mediterranean Diet and a key contributor to its cardioprotective effects. Extra-virgin olive oil is a source of oleic acid and contains phenolic compounds, including hydroxytyrosol and oleuropein, which provide olive oil's characteristic flavor and high stability. Daily consumption of olive oil seems to modulate cytokines and inflammatory markers related to coronary artery disease (CAD) in individuals at risk for cardiovascular diseases (CVD), but clinical trials that evaluate the effect of olive oil and its phenolic compounds on individuals with CAD are sparse.

Omega-3 Fatty Acids

Long-chain polyunsaturated fatty acids (PUFAs) like linolenic acid, EPA, DPA, and DHA are crucial for health. Linolenic acid, found in canola, soybean, and flaxseed oils, nuts, and soybeans, converts to EPA and DHA in the body. Fish and shellfish are rich sources of omega-3 (n-3) fatty acids, often lacking in Western diets.

N-3 fatty acids regulate clotting, lower lipids, inhibit platelet aggregation, and support nerve and retinal membranes, blood pressure, immune function, growth, and brain development. While higher blood n-3 levels correlate with increased prostate cancer risk in some studies, regular seafood consumption is linked to lower prostate cancer death rates, and overall dietary n-3 intake doesn't raise cancer risk.

Studies suggest n-3 PUFAs from fish may reduce depression risk, particularly severe post-partum depression. They also protect against heart disease, stroke, and inflammatory conditions like arthritis and asthma. High-dose n-3s lower triglycerides and improve insulin sensitivity. Maternal DHA intake during pregnancy benefits gestation, birth weight, and child development.

Inadequate n-3 intake can impair neurological development and growth, underscoring their vital role in overall health.

Omega-6 Fatty Acids

Long-chain polyunsaturated fatty acids (PUFAs) essential for health include linoleic acid and arachidonic acid (AA). Linoleic acid, found in soybean, safflower, sunflower, or corn oils, nuts, and seeds, is converted by the body into AA. AA serves as a precursor for eicosanoids like prostaglandins, leukotrienes, and thromboxanes.

Western diets typically contain excessive omega-6 (n-6) fatty acids and insufficient omega-3 (n-3) fatty acids. A deficiency in linoleic acid can lead to poor growth, rough skin, and dermatitis.

High levels of circulating linoleic acid were linked to lower total and coronary heart disease (CHD) mortality in older adults according to the Cardiovascular Health Study. However, a high ratio of n-6 to n-3 PUFAs may promote inflammation and contribute to adverse health effects.

Onion

Bulb vegetable that is a source of the mineral sulfur and, in red onions, anthocyanins (that imparts a red or purple color to certain varieties) and flavonols such as quercetin, which is responsible for the yellow and brown skins of many other varieties. Onion is used to treat CVDs in traditional medicine. Onion exhibited antiplatelet aggregatory, hypocholesterolemic, hypolipidemic, antihypertensive, antidiabetic, anti-hyperhomocysteinemic, antimicrobial, antioxidant, anticarcinogenic, antimutagenic, antiasthmatic, immunomodulatory, and prebiotic effects in laboratory studies. The Netherlands Cohort Study, a large-scale prospective cohort study of diet and cancer (n = 120,852 men and women, aged 55–69 years), found no association between the consumption of onions and leek and the incidence of male and female colon and rectum carcinoma.

Orange, Sweet

More commonly known as the navel orange, popular citrus fruit that is high in vitamin C (83 mg, 92% DV), a good source of folate (48 µg, 12% DV), and supplies pectin and flavonoids.

Flavonoids are potent antioxidants with properties similar to antithrombotic, antifibrotic, and anti-inflammatory drugs. Evidence suggests that regular consumption of flavonoids can enhance peripheral vasodilation and increase blood flow in specific brain regions shortly after eating. This is associated with cognitive benefits.

In a small randomized, double-blind, placebo-controlled study with 22 healthy males aged 30–65, drinking 240 mL of flavonoid-rich orange juice significantly improved cognitive performance compared to a calorie-controlled placebo.

Another study involving 210 participants found that drinking 500 mL of orange juice daily for 12 weeks protected against DNA damage and lipid peroxidation. It also reduced body weight in overweight or obese adults, lowered apoB concentration (a component of LDL cholesterol), and significantly decreased diastolic and systolic blood pressure. Glucose and insulin levels remained within normal range throughout the study.

Oregano

A source of numerous phytochemicals including phenolics, caffeic acid, and rosmarinic acid. In laboratory studies, oregano exhibited antifungal and antioxidant properties. Carnosol, a phytochemical found in common herbs including oregano (as well as in rosemary, sage, and parsley) exhibited antioxidant, anti-inflammatory, and chemoprotective properties in laboratory studies.

Orzo pasta

Primarily a source of B vitamins, particularly B1 (thiamine), B2 (riboflavin), B3 (niacin), and B9 (folate). It also provides small amounts of vitamin E and vitamin K. While specific scientific studies focusing solely on orzo are limited, its nutritional profile indicates potential benefits for energy metabolism (due to B vitamins) and bone health (due to vitamin K). The whole grain varieties of orzo can also contribute to fiber intake, supporting digestive health.

Oyster mushrooms

Good source of various B vitamins, including B2 (riboflavin), B3 (niacin), B5 (pantothenic acid), and B9 (folate). They also contain small amounts of vitamin C and vitamin D when exposed to sunlight. Research suggests that oyster mushrooms possess antioxidant properties due to their phenolic compounds and polysaccharides, which may help in reducing oxidative stress and inflammation in the body. Additionally, they are studied for their potential anti-inflammatory and immune-modulating effects.

Most Common 10 Occlusives:

1. Petroleum Jelly (Vaseline): This is a common occlusive used in skincare to create a barrier that prevents moisture loss.
2. Mineral Oil: Another effective occlusive that forms a barrier on the skin to prevent water evaporation.
3. Shea Butter: Shea butter is rich in fats and oils that form a protective layer on the skin, helping to retain moisture.
4. Beeswax: Beeswax has natural emollient and occlusive properties, making it useful in skincare products.
5. Dimethicone: A silicone-based occlusive agent commonly

used in skincare products to form a barrier on the skin.

6. Plant Oils (e.g., Coconut Oil, Olive Oil): Many plant-based oils can act as occlusives by creating a protective layer on the skin.
7. Cocoa Butter: Cocoa butter is rich in fatty acids that help form a barrier on the skin to retain moisture.
8. Squalane: A lightweight oil that mimics the skin's natural oils and can act as an occlusive to prevent moisture loss.
9. Lanolin: Derived from sheep's wool, lanolin is a natural emollient and occlusive used in skincare for its moisturizing properties.
10. Castor Oil: Castor oil is thick and viscous, making it effective as an occlusive to lock in moisture.

Oligosaccaride
Any of a number of compounds that are made of a few single sugars, such as sucrose or table sugar.

Oxygen
The eighth element in the periodic table. Oxygen has the symbol O and has 8 protons, 8 neutrons, and 8 electrons.

Organelle
A specialized structure within a cell; the chloroplasts in plant cells or mitochondria inside animal cells.

Oxidize
To combine chemically with oxygen.

Palm

Canned heart of palm is high in sodium (622 mg, 27% DV); it

is a good source of fiber (3.5 g, 12.5% DV), folate (57 μg, 14% DV), vitamin C (11.5 mg, 13% DV), and iron (4.5 mg, 25% DV), and contains less than 1 g of total fat from equal parts saturated and polyunsaturated fat. Palm oil, a saturated "tropical" oil, is a source of vitamin E (2 mg, 13% DV), and a common processed foods ingredient. One cup of canned hearts of palm contains 4.5 g of nonheme iron whose absorption is increased by its vitamin C content. Vitamin E in foods is a potent suppressor of cholesterol biosynthesis. Purer Palm oils are known as Batana or Oja.

Papaya

One cup of papaya is an excellent source of vitamin C (83 mg, 92% DV) and vitamin A (1378 IU, 45% DV); and a good source of folate (54 μg, 13% DV); it also supplies fiber (2.5 g, 8% DV) and potassium (264 mg, 7% DV). It is rich in the carotenoid betacryptoxanthin. In traditional and present-day African medicine, papaya is mashed and topically applied to burn wounds to prevent infection, deslough necrotic tissue, and help granulate wounds, effects that are attributed to its proteolytic enzymes and antimicrobial activity. Papaya exerted anti-inflammatory effects in human cells. In vitro research demonstrated papaya's ability to scavenge free radicals and bacteriostatic properties against enteropathogens including E. coli, Salmonella typhi, Staphylococcus aureus, Proteus vulgaris, and Klebsiella pneumoniae.

Paprika

Contains flavonoids and carot- enoids including beta-carotene, lutein, and xeaxanthin, and capsanthin. In Japanese epidemiological studies, serum β-cryptoxanthin was inversely associated with risks of atherosclerosis, Type II diabetes, liver dysfunction, and osteoporosis; and daily intake of β-cryptoxanthin has been inversely correlated with rheumatoid arthritis morbidity.

Parsley

Parsley is an excellent source of vitamin C (19 mg, 21% DV per 1/2 cup [15 g]) and contains flavones. In traditional medicine, parsley is used as an emenagogue. Parsley flavones were anti-inflammatory and parsley constituents were anticarcinogenic in experimental research studies.

Passion Fruit

A source of terpenoids, flavonoids, and fiber. Passion fruit has been used as a folk medicine for its sedative and antihypertensive effects, and for treating anxiety and nervousness. In a laboratory study, Passiflora edulis extract exhibited anxiolytic-like activity. Another laboratory study found that a triterpenoid constituent of Passiflora edulis Sims possessed antidepressant-like activity in animals.

Parsnip

Parsnips are a good source of vitamin C, vitamin K, and several B vitamins including B5 (pantothenic acid) and B9 (folate).They are rich in dietary fiber, promoting digestive health and supporting a healthy gut microbiome. Parsnips also contain antioxidants like vitamin C, which boosts immunity and helps protect against oxidative damage. Additionally, they provide essential minerals such as potassium, beneficial for heart health and muscle function.

Peach

Peaches are a good source of vitamin C (9.9 mg, 11% DV) and contain phytochemicals including chlorogenic acid, phenolics, anthocyanins, and flavonoids. When consumed whole, with the skin intact, peaches supply 2.6 g fiber (9% DV per 1 large fruit). Consuming at least 2 servings of peaches/nectarines per week was associated with a lower risk of estrogen receptor negative breast cancer among postmenopausal women compared with nonusers. Chlorogenic acid protected cells from oxidative damage in a laboratory study.

Peanut

High-protein, low-saturated fat, high-monounsaturated fat legume that is a source of vitamin E (0.7 mg, 5% DV) and resveratrol. There is moderate evidence that peanut and other nut consumption improves serum lipid levels.

Pear

White-fleshed Rosaceae family fruit that is a good source of fiber (5–6 g, 19% DV per medium pear), and flavonoids, such as flavon-3-ols. In rodents, Pyrus communis exerted strong antioxidant activity and protective effects against ethanol or hydrochloric acid induced gastric ulcers. Flavonols exert cardioprotective and anticarcinogenic properties in vitro and in vivo.

Pepper, chili

Containing pungent capsaicinoid compounds, such as capsaicin, that produces burning sensation upon contact with the skin, and dihydrocapsiate. Chili pepper is used as an herbal medicine to treat microbial infections. Capsaicin causes irritation, burning, and sweating, followed by a cooling of body temperature through the evaporation of sweat.

Pepper, sweet bell

An excellent source of vitamins A (720 IU, 24% DV) and C (29 mg, 32% DV), and supplies carotenoids and polyphenolic compounds. Red pepper had higher vitamin C content than green and yellow peppers. Red and yellow peppers showed higher carotenoid content than green peppers. Red pepper showed a higher level of free radical scavenging activity compared to green and yellow peppers.

Peppermint

Cross between water mint and spearmint that contains the monoterpene menthol responsible for its sweet fragrance and cooling sensation. A meta-analysis of 56 randomized controlled trials comparing bulking agents, antispasmodics, or antidepressants with pla- cebo (n=3,725 IBS patients aged over 12 years) found a beneficial effect for antispasmodics over placebo for improvement of abdominal pain. Subgroup analyses for different types of antispasmodics found statistically significant benefit for peppermint oil.

Persimmon

It resembles an orange tomato in appearance (size, color, firmness, and the calyx or stem at the top center of the fruit) and is a source of dietary fiber and polyphenols. Tannin, a type of soluble fiber, in persimmon bound bile acids in vitro. Both fresh and dried persimmon possess high amounts of bioactive compounds that have antioxidant and free radical scavenging potential.

Phenolic Compounds

Epidemiological studies suggest that polyphenols exert cardioprotective effects and that consumption of fruits and vegetables is associated with a reduced incidence of cardiovascular disease, in addition, to reduced risk of developing diabetes, cancer, and stroke, partly due to phenolic compounds. Phenolic compounds have exerted antioxidant, anti-inflammatory, vasodilatory, antifibrotic, apoptotic, antimicrobial, and metabolic effects in laboratory studies.

Major phytochemicals found in plant foods that include:

• phenolic acids such as:
• hydroxycinnamic acids (e.g., chlorogenic acid, p-coumaric acid,

ferulic acid).
• hydroxybenzoic acids (e.g., gallic acid, ellagic acid).

• flavonoidssuchas:
• flavonols (e.g., quercetin, myricetin, kaempferol).
• flavanols/flavans (e.g., catechin, epicatechin, proanthocyanidins). • flavanones (e.g., hesperitin, naringenin, eriodictyol).
• flavanes (e.g., chrysin).
• flavones (e.g., apigenin, luteolin).
• isoflavones/isoflavonoids (e.g., daidzein, genistein, glycitein).
• anthocyanins (e.g., cyanidin, malvidin, petunidin).

Phosphorus

Major mineral whose major functions include energy transfer within the adenosine triphosphate molecule, bone structure as a component of hydroxyapatite crystal, pH maintenance, and structural component of cell phospholipid membranes. Phosphorus deficiency resulting in hypophosphatemia causes cellular dysfunction, symptoms of which include anorexia, anemia, muscle weakness, bone pain, rickets, osteomalacia, general debility, increased susceptibility to infection, paresthesias, ataxia, confusion, and even death.

Phytate

The primary storage compound of phosphorus in seeds. It binds to calcium, iron, potassium, magnesium, manganese, and zinc in the gut and renders them unabsorbable. Phytates are found in legumes, seeds, and the husks of whole grains. Phytic acid is an antioxidant.

Phytoestrogens

When circulating estrogen level is low, phytoestrogens are thought to become estrogenic; when circulating estrogen level is are high, phytoestrogens become anti-estrogenic. The effect of phytoestrogens on the frequency and/or severity of hot flushes (hot flashes) in perimenopausal or postmenopausal women experiencing menopausal symptoms found significantly greater reduction in hot flush frequency compared to placebo, using isoflavone supplements.

Pineapple

An excellent source of vitamin C, pineapple supplies vitamin A, and all B vitamins except vitamin B12. Pineapple is a natural source of the plant protease bromelain that is used orally as an anti-inflammatory agent. Pineapple is a source of papain, a proteolytic enzyme. In vitro and in vivo studies demonstrate that bromelain exhibits various fibrinolytic, antiedematous, antithrombotic, and anti-inflammatory activities, but that there is insufficient evidence to rate its effectiveness for anti-inflammation or other uses in humans. Bromelain also contains chemicals that interfere with the growth of tumor cells and slow blood clotting. Bromelain and papain enhance bioflavonoid absorption.

Pine Nuts

Pine nuts provide vitamins E, K, B1, and B3, along with healthy fats and antioxidants. Scientific findings support their potential health benefits, including antioxidant protection, heart health support, and potential weight management benefits. Including pine nuts in your diet as part of a varied and balanced eating plan can contribute to overall health and wellbeing.

Pistachio

Nut eaten raw or roasted that is a source of protein, vitamin E, and the antioxidant lutein. Antioxidant levels are higher in raw pistachios than in roasted pistachios and in natural shells versus in chemically bleached shells. Records show that people ate pistachio nuts as far back as 7000 BC. As part of an antiatherogenic diet, pistachios may contribute to a reduction in serum oxidized LDL. A meta-analysis of 21 randomized, controlled trials conducted to evaluate the effect of nut consumption on blood pres- sure as primary or secondary outcomes in adult populations aged≥18 y found that total nut consumption lowered systolic blood pressure in participants without Type II diabetes, and that pistachios had the strongest effect on reducing systolic and diastolic blood pressure.

Plantain

It supplies 65 g of carbohydrate and per 1 cup; in addition, it is an excellent source of vitamin A (2227 IU, 74% DV) and potassium (857 mg, 24% DV), and is a good source magnesium (76 mg, 17% DV) and fiber (5 g, 18% DV), while also supplying 9% DV of vitamin E (2.79 mg). Plantains are also a source of flavonoids. A laboratory study found that plantain soluble fiber prevented E. coli translocation in Crohn's Disease cells which may have implications in preventing
Crohn's Disease by limiting harmful bacteria's access to vulnerable intestinal cells.

Plum

An average, 2-inch-diameter fresh plum supplies approximately 7.5 g of carbohydrate and 1 g of fiber, in addition to vitamin C, vitamin E, beta-carotene, and phytochemicals such as phenolics, quercetin, myricetin, and kaempferol. A 100-g serving (10 prunes or 3/4 cups) of dried prunes provides approximately 6.1 g of dietary fiber, 2–3 mg of boron, and 745 mg of potassium, making them a good source of fiber, and an excellent source of potassium. A laboratory analysis found that phenolic compounds in prunes, pitted prunes, and prune juice inhibited LDL oxidation in vitro. Animal and limited human data suggest a role for dried plum in preventing bone loss.

Polyunsaturated Fatty Acids

Oils that include omega-3-fatty acids (n-3) that are sourced from plants and fish, and omega-6 fatty acids (n-6) that are sourced from plants and have long chains of fatty acid chains and contain more than one double bond. Meta-analyses of randomized, clinical trials found no beneficial effects of LCPUFA supple- mentation on the physical, visual and neurodevelopmental outcomes of infants born at term.

Polyunsaturated fatty acid intake significantly decreases the risk of cardiovascular disease and has also been shown to decrease the risk of Type II diabetes.

Pomegranate

Pomegranate is high in vitamin C and fiber as well as phytochemicals including ellagic acid, ellagitannins, punicic acid, anthocyanidins, anthocyanins, estrogenic flavonols, and flavonoids. In traditional medicine, Punica granatum has been used as an antiatherogenic, antidiarrheal, and to treat parasitic and microbial infections, ulcers, aphthae (mouth inflammations), hemorrhage, and respiratory complications. Pomegranate constituents, such as ellagic acid, exerted antioxidant properties in experimental research.

Potassium

Potassium deficiency affects neural transmission, muscle contraction, blood pressure, and vascular tone. Dietary potassium has significantly lowered blood pressure in both hypertensive and nonhypertensive patients in observational studies, clinical trials, and several meta-analyses.

Potato, sweet

One cup of baked sweet potato with its skin is an excellent source of potassium, supplying 950 mg of potassium (30% DV), and, if the skin is consumed, it is a good source of fiber; also, a source of phenolics and flavonoids. One baked sweet potato (3 and a 1/2 oz serving) provides over 8,800 IU of vitamin A (more than 100% DV), calories, is rich in complex carbohydrates, an excellent source of vitamin C, and a good source of iron and fiber. Constituents in sweet potato exerted antioxidant, antiradical, and antiproliferative effects in laboratory studies.

Potato, white

Tuber vegetable rich in carbohydrate, resistant starch, vitamin C, potassium, and, when its skin is consumed, fiber.Some of the phytochemicals in white potatoes include glycoalkaloids (which can exert beneficial or harmful effects), aglycones, and phenolic compounds. In laboratory studies, solanine, a glycoalkaloid, was found to possess anticarcinogenic and anti- prostate-cancer effects, but solanine should not be consumed. Potato peel contains phenolics that exert antioxidant effects.

Prebiotics

Nondigestible dietary constituents (Fiber) that selectively stimulate the growth and/or activity of beneficial microorganisms in the large intestine. Prebiotics beneficially modify gut microbial balance.

Prickly pear

Prickly pear contains vitamin C, fiber, flavonoids, and carotenoids. In Mexico, the prickly pear cultivar, Opuntia streptacantha, is traditionally used in the treatment of diabetes mellitus. According to a review, there is some preliminary clinical evidence that a single dose of the specific species of prickly pear cactus called Opuntia streptacantha can decrease blood glucose levels by 17–46% in patients with Type II diabetes, but it is not known if extended daily use can consistently lower blood glucose levels and decrease HbA1C levels. Opuntia streptacantha juice was confirmed by maltose tolerance test to be an antihyperglycemic agent.

Probiotics

Live microorganisms that, when consumed in adequate amounts, confer a health benefit. Probiotics are used to treat diarrhea, urinary tract and female genitourinary tract infections, irritable bowel syndrome, pouchitis following surgical removal of the colon, and to prevent and manage atopic dermatitis (eczema) in children. Probiotic bacteria mechanisms of action include that they produce butyric acid that neutralizes the activity of dietary carcinogens; attach to enterocytes inhibiting the binding of enteric pathogens and initiating signaling of events that result in the synthesis of cytokines; and exert an influence on commensal microorganisms in the gastrointestinal tract.

Protein

Macronutrient whose constituent amino acids are used to synthesize body proteins or are metabolized as a source of energy. Protein consumption has minimal influence on glycemic response and insulin requirement.

Psyllium

Psyllium seed husk is made into cereals and is a chief ingredient in bulk laxatives. Psyllium seed husk contains soluble fiber. Psyllium is an effective laxative and stool softener that may be effective in relieving the symptoms of irritable bowel syndrome, including diarrhea; reducing high blood glucose levels in individuals with Type I and II diabetes. Small clinical trials have shown psyllium fiber to be associated with lower mean daily glucose concentrations, lower post-meal glucose concentrations, fewer hypoglycemic events, lower hemoglobin A1C levels, and lower insulin concentrations in people with diabetes mellitus.

Pummelo

Its segments are white- or red-fleshed. Pummelos are excellent sources of vitamin C and contain flavonoids. Fresh red pummelo juice is an excellent source of antioxidant compounds that in a laboratory study scavenged superoxide anion free radicals and hydrogen peroxide radicals. Citrus genus fruits contain limonoids, bitter-flavored secondary metabolites that have exhibited antibacterial, antifungal, antimalarial, anticancer, and antiviral properties in laboratory studies.

Pumpkin

Squash family vegetable that is an excellent source of vitamin A (7050 IU per 1/2 cup, 235% DV) from beta-carotene, and notably is rich in other carotenoids. A review examining natural treatments for benign prostatic hypertrophy found no convincing evidence to support the use of Curcubita pepo alone for its treatment. In experimental research, pumpkin exhibited moderate antioxidant activity, and moderate to high alphaglucosidase and angiotensin-converting enzyme inhibitory activities, which may have implications for hyperglycemia and hypertension management, respectively. Pumpkin consumption was inversely associated with the development of lung cancer in a case-control study comparing the dietary patterns of subjects with incident lung cancer (n = 371) to controls (n = 496); neither cases nor controls had a neoplastic history.

Pumpkin seed

Pumpkin seeds, (roasted per 1 oz), are an excellent source of vitamin E (3.72 mg, 25% DV), phosphorus (333 mg, 26% DV) and zinc (2.17 mg, 20% DV), and contain iron (2.7 mg, 10% DV) and fiber (2 g, 7% DV). Pumpkin seeds 'vitamin E is the gamma-tocopherol form (a type of natural vitamin E).1 In a small study (n=20 boys age 2–7 years), pumpkin seed treatment as compared with treatment with orthophosphate (the control compound),reduced calcium oxalate bladder stone formation.

Peptide bond
The attachment between two amino acids connected through the carboxylic acid group and the amine group.

Periodic Table of Elements
A chart categorizing all of the chemical elements.

PH
A term used to describe the acidity or basicity of a solution.

PH meter
An electronic meter that detects and displays the pH of a solution.

Phenolphthalein
An acid-base indicator; above pH 9 it is pink, below pH 9 it is clear.

Plastic
Any substance made of polyethylene or other similar polymers.

Plot
The graphical representation of data.

Polyethylene
A polymer chain that makes many different plastics.

Polymer
A general term for any chain of repeating molecular units.

Polypeptide
A chain of many peptide units, usually less than a complete protein.

Polysaccharide
Any of a number of compounds that are made of many single sugars such as starches and cellulose.

Pores
The small holes that are in a filter.

Precipitate
The products of a chemical reaction that are not able to dissolve in the remaining solution and can look like mud/sand/ or snow.

Protein
A polymer of amino acids.

Protons
One of the three fundamental particles that make atoms. It has an atomic mass of 1 and carries a positive charge.

Parasite
An organism that lives in or on another different living thing, benefitting at the hosts expense.

Pasteurize
To kill microbes through heat. Many foods, such as milk and wine, are pasteurized.

Pathogen
A microorganism that can cause disease.

Photosynthesis
The ability of a living thing to make the essential chemicals for life using energy from sunlight.

Polymer
A substance made from many long molecules of the same kind. Most plastics are polymers.

Prokaryote
A living thing whose cells have no central nucleus. All prokaryotes are single celled microorganisms.

Protein
Nonliving chemical that carries out essential life processes inside living things. Enzymes and RNA are examples of proteins.

Photovoltaic
The production of electricity by light.

Probability cloud
In quantum mechanics, the position of a particle cannot be determined precisely. Instead there is a fuzzy region of places inside which there is a chance that it might turn up that region is a probability cloud

Paradox
A puzzle in which the pieces don't seem to fit together, or in which the answer seems to contain a contradiction

Quercetin

Major flavonoid in the US diet consumed in daily amounts of 4 or 5 mg through plant foods, such as white fruits and vegetables, for example, apples, pears, and onions, but in pigmented plant foods as well, for example, red leaf lettuce, leafy green vegetables, green tea, quinoa, wine, and asparagus; whereas median daily intake of quercetin in Japan, because it may consume a higher proportion of calories from plant foods, is estimated to be 15.5 mg/day. The potential to lower disease risk factors, but human data are limited. Quercetin has been shown to modify eicosanoid biosynthesis; protect LDL from oxidation; prevent platelet aggregation; and relax cardiovascular smooth muscle properties which may have antihypertensive and antiarrhythmic effects; and to exert antiviral and carcinostatic properties. Another case-control study (n=582 patients with incident lung cancer and n=582 age, sex, and ethnicity matched control subjects) found a statistically significant inverse association between lung cancer risk and quercetin, from foods such as onions and apples, with onions being particularly strong against squamous cell carcinoma.

Quail

Quail is a lean source of protein, rich in essential vitamins such as B vitamins (particularly B12), vitamin A, and minerals like iron and zinc.Quail meat is known for its high protein content, which is essential for muscle growth and repair. It also contains significant amounts of vitamin B12, crucial for nerve function and red blood cell production. Quail meat is lean and low in fat, making it a healthier protein option compared to many other meats.

Quince

Quince pulp, peel, and jam exhibited antioxidant activity in a laboratory study. Whole Cydonia oblonga Miller fruit, including pulp, peel, and seed, exhibited antiproliferative properties against human kidney cancer cells and colon cancer cells in a laboratory study.

Quinoa

Quinoa contains fiber (2.6 g, 9% DV), zinc (1 mg, 9% DV),

selenium (2.6 µg, 5% DV), and vitamin E (0.585 mg, 4% DV); and is good source of magnesium (59 g, 14% DV) and phosphorus (140 mg, 11% DV) per 1/2 cup; in addition, quinoa supplies phytosterols and flavonoids. Quinoa's content of amino acids, though higher than that of many common grains, does not include adequate amounts of the essential amino acids threonine, lysine, and phenylalanine; hence, it is not a complete protein, as has been claimed, because it does not contain all essential amino acids necessary for human growth. Quinoa seeds fed to rats reduced serum total cholesterol, glucose, LDL, and triglycerides.

Queen Garnet Plum

The Queen Garnet plum is a variety known for its exceptionally high levels of anthocyanins, which are powerful antioxidants. It also contains vitamins A, C, and K, as well as minerals such as potassium.Research has shown that the Queen Garnet plum has significant antioxidant activity, which may help protect cells from oxidative stress and inflammation. Studies suggest potential benefits for cardiovascular health, skin health, and immune function due to its high anthocyanin content.

Quantum theory
A strange branch of physics that shows how light can act like a particle and how electrons and other particles have wavelike properties.

Quorum sensing
A system based on signal chemicals and receptors used by bacteria to sense their numbers and coordinate their actions

Quantum interference

All waves can collide and become mixed up with one another. Quantum interference is the name given to this occurrence when it happens with subatomic particles, owing to their wavelike nature.

Radicchio

Radicchio, per 1 cup, contains folate (24 µg, 6% DV) and is an excellent source of vitamin K (102 µg, 25.5% DV). Phytochemical constituents in radicchio include lutein + xeaxanthin (3,533 µg [3.5 mg]), and phenolics. In a laboratory study, Cichorium intybus exhibited antioxidant properties.

Radish

Radishes are so low in calories, they are considered a calorie-free food and they are a source of vitamin C. Traditional uses of radish include to treat peptic disorders, bile duct dyskinesia, loss of appe- tite, inflammation of the mouth and pharynx, bronchitis, fever, colds, and cough, and to prevent infection or inflammation or excessive mucus of the respiratory tract.

Raisin

Raisins are a good source of potassium (749 mg, 11% DV), and fiber (3.7 g of both soluble and insoluble fiber, 13% DV) per 50 g or approximately ⅓ cups. Compared to grapes, raisins contain more calories (approximately 149 calories per 1/3 cup) and fructans (see: Fructan), but fewer phenolic compounds that are lost during processing, including procyanidins and flavan-3-ols. In a crossover-design study (n=16 healthy adults), 84 g (0.58 cup or slightly more than 1/2 cup11) of sun-dried raisins improved measures of colon health, including intestinal transit time, fecal weight, and fecal bile acid composition.

Raspberry

High-vitamin C berry that provides 4 g of fiber per 1/2 cup serving (14% DV), phenolics, anthocyanins, and caffeic acid (not to be confused with caffeine—the two substances are unrelated). In laboratory studies, caffeic acid exhibited antioxidant, immunomodulatory, and anti-inflammatory properties. In an experimen- tal model, caffeic acid protected human skin from photo-oxidative damage.

Rambutan

Rambutan is a tropical fruit rich in vitamin C, fiber, and antioxidants. It also provides small amounts of iron, calcium, and vitamin A. Research suggests that rambutan contains antioxidants such as flavonoids and polyphenols, which may help protect cells from damage caused by free radicals. These antioxidants may also contribute to anti-inflammatory and antimicrobial properties.

Resveratrol

Polyphenol linked to the "French Paradox" phenomenon, referring to the anti-atherogenicity of the high-saturated-fat French Diet, presumably due to the inclusion of red wine. Red wine is the major dietary source of resveratrol, while white wine has a low content of resveratrol, a phytochemical that is particularly rich in grape skins and also found in peanuts, as well as other foods. Resveratrol inhibited LDL oxidation and platelet aggregation, and exerted anti-inflammatory and anti-proliferative effects in laboratory resea rch. A second meta-analysis of 15 RCTs (n=658) concluded resveratrol supplements might not be able to change concentrations of the inflammatory markers interleukin-6 or tumor necrosis factor-α, but found a "possible decreasing effect of resve- ratrol on CRP."

Red Cabbage

Red cabbage is rich in vitamins C, K, and B6. It also contains dietary fiber, manganese, and antioxidants such as anthocyanins, which give it its vibrant color. Research suggests that red cabbage has anti-inflammatory and antioxidant properties, which may help in reducing the risk of chronic diseases. It is also studied for its potential role in promoting heart health and supporting digestion.

Rhubarb

Rhubarb is used to make pies, jam, jelly, and sauces, and is an excellent source of vitamin K (40 µg, 33% DV) and a good source of mostly insoluble but also soluble fiber totaling 2.5 g of total fiber (9% DV) per 1
cup. Thought it contains 344 mg of calcium (26% DV) per 1 cup, its high oxalic acid content somewhat blocks calcium absorption. It is also a source of the phytochemical anthraquinone. Rhubarb has been used in traditional Chinese medicine to treat constipation, blood stasis, diabetic nephropathy, chronic renal failure, acute pancreatitis, gastrointestinal bleeding, and other diseases. Anthraquinone exhibited antiviral properties in laboratory studies. Anthraquinones have cathartic effects in vivo.

Riboflavin

Also referred to as vitamin B2. Water-soluble vitamin that functions as a coenzyme in redox reactions. The best sources of riboflavin are milk and enriched grain products. The median riboflavin intake in the US meets the riboflavin RDA. Deficiency symptoms of riboflavin, characterized by low riboflavin levels in the blood (ariboflavinosis) include: sore throat, edema of the pharyngeal and oral mucosa, angular stomatitis, magenta tongue, and impaired metabolism of vitamin B6. Higher intake of riboflavin was associated with a reduction in age-related lens opacification progression in a subset (n = 408 aged 51–72 y at baseline) of The Nurses 'Health Study.

Rosemary

Contains rosmarinic acid, "one of the most important and well known natural anti- oxidant compounds." In experimental studies, rosemary exhibited anti-inflammatory, antispas-modic, analgesic, antirheumatic, carminative, cholagogue, diuretic, expec- torant, antiepileptic, and neuroprotective, properties.

Rutabaga

Cooked rutabaga is a good source of vitamin C (16 mg, 17% DV), and rutabaga supplies potassium (184 mg, 5% DV) and glucosinolates.

"R" group

The group of atoms in ana amino acid that makes it unique. The "R" group can be anything from a simple hydrogen atom to a complicated ring of atoms

Radioactive dating - A way of figuring out the age of rocks, fossils, bones, and other ancient materials by calculating the number of radioactive atoms that have decayed.

Ring of Fire

A circle of intense earthquake activity around the edges of the Pacific Ocean, caused by the movement of Earth's tectonic plates.

Radiation

Strictly called radioactivity, this involves energetic particles or a zap of energy that comes shooting out of a atomic nucleus when it breaks apart. Radioactivity can damage DNA.

RNA

An important life chemical that carries out many functions in the cell, including copying DNA. It stands for ribonucleic acid.

Radioactive

Describes an unstable substance with an atomic nucleus that breaks down and releases nuclear radiation.

Saffron

While saffron is not a significant source of vitamins, it contains potent bioactive compounds that contribute to its antioxidant properties and potential health benefits. Scientific findings support its traditional uses in promoting mood, neuroprotection, and possibly cancer prevention. However, more research is needed to fully understand its mechanisms of action and to validate its therapeutic effects in humans.

Sage

Earthy-flavored herb used fresh or dried to flavor meats, fish, and stews. Consumed as a culinary herb in small quantities, sage is not an appreciable source of nutrients. In traditional medicine, sage tea was used to treat dyspepsia, mouth and throat inflammation, excessive sweating, and minor skin inflammations. Ancient Greek physicians used a sage and water solution to stop wounds from bleeding and to clean sores and ulcers. In Native American rituals, dried sage is burned to promote healing, wisdom, protection, and longevity. Many species of sage have been used historically as "brain enhancing tonics." Salvia officinalis L. (sage) has been widely used in Tunisian traditional medicine, and Salvia species in European folk medicine, for the treatment of memory disorders. In laboratory research, S. lavandulaefolia and/or its constituents have demonstrated anticholinesterase, antioxidant, anti- inflammatory, estrogenic, and central nervous system depressant, or sedative, effects. There is insufficient scientific evidence to support sage as a treatment for a sore throat. S. officinalis demonstrated anticholinesterase (i.e., acetylcholinesterase inhibitory/acetylcholine-binding/ cholinergic binding), antioxidant, anti- inflammatory, and estrogenic (i.e., anti-hot flush [also referred to as hot flash]) properties in vitro. Sage purportedly reduces lactation and has been used to aid with weaning or an overabundant milk supply, but studies evaluating the effect of sage and human lactation could not be found in the scientific literature.

Saponin

Glycoside compound found in a wide variety of plants, such as soybeans, chickpeas, peanuts, spinach, quinoa, and beer. When fruits and vegetables are cut or otherwise processed, saponin produces a soapy foam. Dietary saponins, either isolated or as saponin containing food plants, have reduced plasma cholesterol levels in animal studies. Experimental data show saponin exert blood-pressure lowering effects. Saponins demonstrated immune properties in vitro including stimulating phagocytic, bactericidal, and adhesion activities of white blood cells.

Selenium

Trace mineral found in meat, seafood, and certain plant foods, such as nuts, especially Brazil nuts, which supply 290 µg of selenium (Se) per 1 nut18 (five times the Reference Dietary Intake amount of 55 µg). Se is a cofactor for antioxidant enzymes, including glutathione peroxidase, whose main role is to prevent oxidative damage. Se is necessary also for thyroid function; immune function; normal testicular development, spermatogenesis, and spermatozoa motility and function in males; and it exhibits insulin-mimetic properties. Se plays a protective role against prostate cancer development and its progression to advanced stages. Researchers concluded that if there is a benefit of Se on prostate cancer risk, it appears to be limited to those who are Se-deficient. Se is a nutrient consumed less by short sleepers (<5 h per night) in an observational study that examined associations between self-reported sleep duration and intake of various dietary components (n = 5587 participants in The 2007–2008 National Health and Nutrition Examination Survey); normal sleep duration was associated with the greatest food variety. Se status appears to have an impact on the development of thyroid pathologies. Adequate Se, together with optimal iodine and iron intake, is required for a healthy and functional thyroid during development, adolescence, adulthood, and aging.

Sesame Seeds

Sesame seeds are rich in essential nutrients such as calcium, iron, magnesium, phosphorus, zinc, and selenium. They also contain fiber and healthy fats. Research indicates that sesame seeds have antioxidant properties and may help lower cholesterol levels and reduce inflammation. They are also studied for their potential benefits in promoting heart health and supporting bone health.

Shallot

Miniature purple onion that contains allyl sulfides and flavonol glycosides, including quercetin. Used for their strong, characteristically burning flavor in salads and salad dressings.

Snap Pea

Fresh shoots of the snap pea are used as microgreens, i.e., young sprouts or "pea shoots," in salads or as a garnish, and dried, split peas are an excellent source of choline (153 mg, 28% DV) per 1/2 cup in addition to many other nutrients. Snap peas contain fiber (2 g, 7% DV) and are an excellent source of vitamin C (36 mg, 40% DV), while also supplying phenolic compounds and flavonoids.

Snow Pea

The snow pea supplies 2 g of fiber (7% DV) and is a good source of vitamin C (15 mg, 16% DV) per 3 oz. Sativin isolated from snow pea (Pisum sativum var. macrocarpon) exerted antifungal activity in vitro against a variety of organisms. In an animal study, diets containing P. sativum reduced serum total cholesterol, LDL, VLDL, and cholesterol-to-HDL ratio.

Sodium

Major mineral electrolyte and primary intracellular cation that is necessary for maintaining fluid balance, blood volume, blood pressure, nerve impulse transmission, acid-base balance, muscle function, and plays many other vital roles. Sodium deficiency causes disturbances in acid-base balance, poor appetite, muscle cramps, confusion, apathy, constipation, and cardiac arrhythmia, Sodium excess is associated with hypertension.

Soybean

Edamame are immature soybeans that are good sources of fiber (5.2 g fiber, 17% DV) and zinc (1.25 mg, 11% DV), and excellent sources of vitamin C (37 mg, 41% DV), calcium (250 mg, 19% DV), potassium (793 mg, 23% DV), iron (4.5 mg, 25% DV), and folate (211 µg, 52% DV) per 1/2 cup. Some evidence suggests soy protein added to a diet low in saturated fat reduces the risk of coronary heart disease, but there are inconsistent findings concerning the ability of soy protein to lower LDL cholesterol.

Spinach

Dark green leaf vegetable that is a good source of folate (58 µg, 14.5% DV) and an excellent source of vitamins A (2,813 IU, 94%) and K (144 µg, 120% DV) per 1 cup. Unlike other dark green vegetables, spinach is a poor calcium source, providing only 30 mg per 1 cup (2% DV). Despite its high-iron reputation, spinach is only a fair source of iron, providing approximately 0.8 mg of nonheme iron per 1 cup (4% DV)55 depending on the brand (some brands may contain slightly more), whereas a "good source" would supply 10–19% DV. Spinach is high-iron vegetables, than many vegetables (e.g., 1 cup of green onions supplies 0.48 mg and 2.6% DV and watercress has 0.7 mg and 0.3% DV of iron), whereas legumes are high-iron vegetables, supplying up to 10 mg per cup (55% DV of iron). Consumption of green leafy vegetables is associated with a reduced risk of several types of cancer such as pancreatic cancer, and cardiovascular disease.

Squash

Squash is high in starch (15 g per 1/2 cup), rich in vitamin A (400 IU, 13% DV), and a source of all B vitamins except vitamin B12. Hubbard squash (Curcubita maxima), is carotenoid-rich and therefore a good source of vitamin A and a significant source of starch (15 g per) per 1/2 cup. Greater intake of yellow/orange vegetables was one of several dietary pat- terns that were associated with significantly lower breast cancer risk in a 30-year observational study.

Stanols

Including plant sterol/stanol esters in the diet helps to lower blood total and LDL cholesterol levels. A meta-analysis that included 14 studies (n=531 patients) showed that there is no statistically or clinically signifi- cant difference between plant sterols and plant stanols in their abilities to modify serum total cholesterol.

Starchy Vegetables

Vegetables that generally have a higher carbohydrate content (~15 g of carbohydrate per ½–1 cup serving) than non-starchy vegetables, such as lettuce and tomatoes (~5 g of carbohydrate per ½–1 cup serving), and therefore are considered to be equivalent to grains for diet planning purposes. Starch is the major glycemic carbohydrate, along with sugars, in foods. Starchy vegetables increase serum glucose. However, some starchy vegetables contain resistant starches that either reduce or have no effect upon glycemic load in addition to having other health benefits—see also: Fructan.

Strawberry

Berry that is a source of fiber (2.3 g, 8% DV), potassium (177 mg, 5% DV) folate (28 μg, 7% DV), and an excellent source of vitamin C (68 mg, 75% DV). Strawberry constituents ameliorated allergy symptoms in a laboratory study. "Strawberry phenolics are best known for their antioxidant and anti-inflammatory action, and possess direct and indirect antimicrobial, anti-allergy and antihypertensive properties, as well as the capacity to inhibit the activities of some physiological enzymes and receptor properties."

Sonic boom
A "thunderclap" when the shock wave from an airplane breaking the sound barrier—traveling faster than the speed of sound—touches the ground.

Subatomic
Anything smaller than an atom.

Soap
A molecule with a charged end (head) and a long uncharged end (tail). Soap can be dissolved in both water and oily substances and is used for cleaning for this reason.

Sodium
The eleventh element on the periodic table with the symbol Na. It has 11 protons, 11 neutrons, and 22 electrons. Sodium is an alkali metal.

Sodium bicarbonate
Baking soda.

Sodium chloride
The molecule that results when sodium atoms and chloride atoms are bonded together. This is also known as table salt.

Solid state
The state of a substance when it is in a solid form, such a ice, and cannot be easily changed in shape.

Solvent
Any substance in which another substance is dissolved.

Spontaneous
Occurs when a chemical reaction proceeds without added energy input.

Starch

Polysaccharides that are found in both plants and animals.

Symbol
A unique symbol given to each chemical element. It is usually the first two letters of the element name.

Semiconductor
A material that conducts electricity better than an insulator, but not as well as a metal. These materials are used to control electric current and are the brains inside of electronic devices.

Spin
A property of subatomic particles that is a little like the momentum an object picks up with it rotates. Spin can take only certain fixed values.

Stable
Any molecule, atom, particle, or system that is not likely to change physically or chemically or decay radioactively.

Standard Model
The complete theory that describes all subatomic particle and the forces that act upon them.

Subatomic Particle
Anything that is smaller than an atom.

Superconductor
A material with zero resistance to electrical currents. It can make a megastrong magnet, but needs to be extremely cold to work.

Superposition
When two or more things happen in the same place; the principle that particles can exist in several different quantum states at once.

Tamarind

A good source of potassium (377 mg, 11% DV) that contains large amounts of the phytochemical tartaric acid. Tamarind has been used in traditional African med- icine as a laxative, for abdominal pain, to treat diarrhea and dysentery, to treat helminth infections, for wound healing, and to treat malaria, and for fever, constipation, inflammation, cell cytotoxicity, gonorrhea, and eye diseases. Tamarindus indica seeds exhibited antioxidant properties and Tamarindus indica extract exhibited antidiabetic properties in laboratory research. Tartaric acid may be beneficial for digestion and bowel regulation.

Tangerine

A large (2¾-inch diameter) tan- gerine supplies 199 mg (6% DV) of potassium, 19 µg (5% DV) of folate, is an excellent source of vitamin C (32 mg, 35% DV), and contains an array of flavonoids (flavones, flavonols, and flavanones, such as kaempferol). Both canned tangerine (mandarin oranges) and fresh tangerines are a signifi- cant source of the antioxidant beta-cryptoxanthin, one of five carotenoids that predominate in human plasma.In a laboratory study, beta-cryptoxanthin exhibited antitumor effects. Kaempferol and some of its glycosides have a wide range of beneficial pharmacological activities, including analgesic, antiallergic, anticancer, antidiabetic, anti-inflammatory, antimicrobial, antiosteoporotic, anti- oxidant, antiapoptotic, anxiolytic, cardioprotective, estrogenic/antiestrogenic, and neuroprotective.

Tannin

Secondary plant metabolite classified as a proanthocyanidin. Found in many foods including black-eyed peas, grapes, lentils, persimmon, black and green teas, coffee, and red and white wines. Tannins contribute color, bitterness, and astringency, for example, of wine. In in vitro studies, tannins exhibited chemoprotective properties and bound bile acids. Tannins can form insoluble complexes with carbohydrates and protein; their concomitant ingestion with nonheme iron inhibits nonheme iron absorption and can significantly impact iron status. Consuming vitamin C improves nonheme iron absorption and nullifies tannins 'opposite effects.

Tarragon

It has been used in tra- ditional folk medicine to treat pain and gastrointestinal disturbances; and in Iranian folkloric medicine, tarragon has been used orally as an antiepileptic remedy. In vitro, tarragon exhibited antioxidant and antinociceptive properties and induced potent anticancer effects, including inhibiting esophageal squamous cell carcinoma. Tarragon exhibited antiseizure and sedative effects that were attributed to its monoterpenoids content in laboratory research.

Taro

Taro is rich in fiber, vitamins C and E, potassium, magnesium, and iron. It is a good source of complex carbohydrates. Studies suggest that taro has potential health benefits, such as improving digestive health due to its high fiber content and providing antioxidant properties that may help protect against chronic diseases.

Tea

One of the most ancient and popular beverages consumed around the world, second only to water that is prepared by steeping Camellia sinensis leaves in water.Tea phytochemicals include alkaloids (caffeine, theophylline, and theobromine), chlorophyll, flavon-3-ols, including catechins, and trace elements, such as fluoride. Green tea is particularly rich in catechins which are the substances most associated with health benefits. Some epidemiological studies comparing tea drinkers to non-tea drinkers support the claim that drinking tea prevents cancer; others do not. Laboratory studies have shown that "tea catechins act as powerful inhibitors of cancer growth in several ways: they scavenge oxidants before cell injuries occur, reduce the incidence and size of chemically induced tumors, and inhibit the growth of tumor cells. In studies of liver, skin, and stomach cancer, chemically induced tumors were shown to decrease in size in mice that were fed green and black tea.

Teff

Teff is a gluten-free grain that is rich in nutrients such as iron, calcium, magnesium, manganese, phosphorus, and zinc. It also provides a good source of protein and dietary fiber.Teff has gained attention for its nutritional benefits, particularly in providing essential minerals and vitamins like vitamin B6. It is also recognized for its potential role in managing blood sugar levels due to its low glycemic index.

Terpenoids

Group of phytochemicals whose main subclasses are monoterpenes (including limonene, carvone, and carveol); diterpenes (including the retinoids); triterpenes (such as saponins); tetraterpenes (e.g., carotenoids, including carotenes and xanthophylls), triterpenoids (e.g., glycosides), and steroids (including stanols, sterols, and tocopherols). Terpenoids exhibited antimicrobial effects in a laboratory study. Cancer prevention advice to eat at least 2.5 cups of vegetables and fruits each day is based on plant foods" 'numerous potentially beneficial bioactive substances, such as terpenes ... that may help prevent cancer."

Thyme

Phenolic compounds have exerted antioxidant, anti-inflammatory, and antimicrobial effects in laboratory studies. Thyme exerted DNA- protective activity in a laboratory study.

Tofu

Tofu is a rich source of soy protein and can be a good source of calcium, depending on whether the soybean is precipitated with a calcium com- pound, such as calcium chloride, but tofu is not a source of vitamin B12, unless it has been fortified with vitamin B12 which would be stated in the ingredients list. Daily soy protein intake has a mild LDL-lowering effect, especially when substituted for animal protein.

Tomatillo

Miniature green tomato-like vegetable used to make salsa verde (green salsa). A medium, 34g tomatillo contains 10 calories, 1 g of fiber (3% DV) and 4 mg vitamin C (4% DV) and is a source of phytochemicals including lutein, beta-carotene, and ixocarpalactone A. Ixocarpalactone A has been shown to be a cancer chemopreventive agent against human cancer cell lines in laboratory studies.

Tomato

Tomato consumption, specifically, tomato foods, cooked tomatoes, and sauces, but not raw tomatoes, were associated with a reduced risk of pros- tate cancer risk in a systematic review and meta-analysis of 30 studies that summarized data from 222 cases and even more participants.

Turmeric

Traditional uses of turmeric include to aid in digestion and liver function, relieve arthritis pain, and regulate menstruation. Curcumin, a vanilloid compound in turmeric, exhibited apoptotic, anti- inflammatory, chemoprotective, antitumor, antioxidant, antiarthritic, anti-amyloid, anti-ischemic, anti-inflammatory, hepatoprotective, and bone-protective effects.

Turnip

One cup of raw turnip is low in calories (36 calories) and carbohydrate (8 g, 2% DV), and high in vitamin C (38 mg, 42% DV). Its greens, commonly boiled or cooked with ingredients such as meats, onions, and garlic, are rich in phenolics and glucosinolates.

TOP 8 FOODS THAT BOOST (T)ESTOSTERONE

1. Legumes

Eating legumes—a family that includes beans, peas, and peanuts —can protect against low testosterone levels and poor testicular function. Legumes are rich in two nutrients associated with higher testosterone levels: zinc and magnesium.

2. Dark, Leafy Greens

Dark, leafy greens are nutrient-dense, high in fiber, and filled with micronutrients that promote health. Some evidence suggests that men who eat more dark, leafy greens have higher testosterone levels than those who do not.

3. RAW Honey

Natural medicine often points to honey's many uses, from soothing a sore throat to its antibacterial properties. More research on humans is still needed to confirm the effect of honey on testosterone.

Research has found that honey can increase testosterone levels in men by:

- Enhancing the viability of Leydig cells
- Increasing the production of luteinizing hormone
- Inhibiting aromatase activity in the testes
- Reducing oxidative damage in Leydig cells

4. Onions

A review published in 2019 found that onions may help male testosterone levels. The study authors found that onions increase the production of luteinizing hormone, enhance the antioxidant defense mechanism in the testes, and defend against inflammation and insulin resistance. More clinical trials are needed to determine how onion can increase testosterone levels in humans.

5. Eggs

Whole eggs contain cholesterol, which your body needs to make testosterone. A study published in 2020 followed young men on a weight training program over 12 weeks. The researchers found that men who ate eggs daily increased their testosterone levels.

The men were separated into two groups: one that ate three whole eggs daily and one that ate six egg whites. The study authors found that the group that ate all the eggs increased their testosterone levels more.

6. Foods with Flavonoids

Eating high levels of flavonoids is good for your health and testosterone production. Flavonoids are a group of compounds in many fruits and vegetables. Research has shown that flavonoids can prevent or delay hypogonadism (decreased functional activity of the gonads) in males as they age.

Foods high in flavonoids include:

- Broccoli
- Kale
- Hot peppers
- Onions
- Rutabagas
- Spinach

7. Oysters

Oysters are full of zinc, a mineral that's essential for making testosterone. Taking in extra zinc may ensure you're not deficient, but getting more than you need won't do much to increase testosterone.

8. Some Herbs

There are countless herbs on the market that tout their testosterone-boosting effects but have little evidence backing them up. A review published in 2021 found that some herbs show promise, although more research is needed.

These herbs include:

- Ashwagandha (root and leaf extract)
- Asian red ginseng
- Fenugreek seed (extract)
- Forskohlii (root extract)

Thought experiment
An imaginary scenario dreamed up by a scientist to test out a theory

Tracer
A radioactive substance that is put inside the body to track the workings of the internal organs.

Thymine
One of the four nucleic acid bases that make DNA.

Titration
The experimental technique used to find out the unknown concentration of an acid or base.

Tungsten
Element number 74 with the chemical symbol W. Tungsten has 74 protons, 74 electrons, and 109 neutrons.

Toxin
A poison or venom produced y a living thing. When produced by microorganisms in the body, it can cause disease.

Udon Noodles

Udon noodles are a type of thick wheat flour noodle used in Japanese cuisine. They provide carbohydrates for energy and a small amount of protein.Studies have shown that consuming traditional Japanese foods, including udon noodles, may contribute to a healthier overall diet due to their balanced nutritional profile and lower fat content compared to some other noodle varieties.

Umami Paste
Umami paste is rich in glutamic acid, which provides the umami flavor. It also contains various amino acids, minerals like potassium, and small amounts of vitamins B and C.Umami, known as the fifth taste, has been studied for its role in enhancing flavor and satiety. It contributes to the overall taste profile of dishes and can influence eating satisfaction.

Ugli fruit

Larger than an orange with yellow-green, easily peeled skin, ugli fruit eaten fresh is an excellent source of vitamin C and fiber.

Uranium
Element number 92 with the chemical symbol U. Uranium has 92 protons, 92 electrons, and 146 neutrons

Unstable
The opposite of stable. A molecule, atom, particle, or system that is likely to change or decay without warning

Vegetables

As a group, vegetables supply folate, beta-carotene, vitamins C, K, and E, magnesium, potassium, and fiber. Different colored vegetables are characteristically rich in certain nutrients, e.g., dark green vegetables are particularly rich in folate, vitamin

K, and magnesium, therefore ChooseMyPlate.gov advice to "Vary your veggies" promotes consuming an array of nutrients and phytochemicals. Each vegetable may contain more than 100 different phytochemicals. Biologic mechanisms whereby vegetables exert healthful effects are likely to be multiple and include both nutrients and phytochemicals. Fruits and vegetables, low in calories because they're high in fiber and water, promote fullness and volume. Vegetables are good substitutes for foods of high energy density.

Vegetable Oil

Vegetable oils such as olive oil, canola oil, or sunflower oil are rich in unsaturated fats, particularly monounsaturated and polyunsaturated fats. They provide vitamin E and phytosterols, which have antioxidant properties. Research indicates that consuming moderate amounts of unsaturated fats from vegetable oils can help lower cholesterol levels and reduce the risk of heart disease. The type and quality of vegetable oil used in cooking can influence its health benefits.

Venison

Venison is a lean meat that is high in protein and low in fat. It provides essential nutrients such as iron, zinc, and B vitamins (particularly B12 and niacin). It also contains omega-3 fatty acids. Venison is often promoted for its lean protein content and beneficial nutrient profile. Studies suggest that consuming lean meats like venison as part of a balanced diet may contribute to overall health and well-being.

Vinegar

In laboratory animals, coumaric acid reduced LDL cholesterol levels and exhibited significant neuroprotective activity following an ischemia- reperfusion injury of the spinal cord. In a 12-week, double-blind, pla- cebo-controlled RCT trial (n = 176 obese subjects), 15 ml of vinegar reduced body weight, body fat mass, and serum triglyceride levels in the treatment group.

Vitamin A

Crucial for vision, corneal health, growth, epithelial integrity, reproduction, immune function, and gene expression, is found in fortified dairy, fruits, and vegetables. It exists as preformed retinol in animal foods like milk and liver, and as provitamin A carotenoids in dark green, orange, and yellow fruits and vegetables, such as spinach and carrots. Deficiency leads to xerophthalmia and other serious conditions; biomarkers include plasma retinol, though liver vitamin A concentration is the gold standard, affected by various health conditions. Excessive intake of preformed vitamin A can harm bone health.

Vitamin B6

Essential for protein, carbohydrate, and lipid metabolism, as well as neurotransmitter synthesis, immune function, and cognitive health, is found in fortified cereals, protein foods, potatoes, and fruits like bananas and watermelon. Deficiency, marked by symptoms such as depression, dermatitis, and anemia, is associated with elevated homocysteine levels and increased risk of mortality. Measurement of B6 status utilizes plasma pyridoxal phosphate levels, with values above 20-30 nmol/L considered normal.

Vitamin B12

Also known as cobalamin, is crucial for folate metabolism, DNA synthesis, red blood cell formation, myelin sheath synthesis, and numerous other functions. Found primarily in animal foods, B12 deficiency can lead to megaloblastic anemia and neurological symptoms such as paresthesias and cognitive decline. Diagnosis relies on markers like serum B12, holotranscobalamin, homocysteine, and methylmalonic acid, with elderly individuals and certain medical conditions predisposing to deficiency despite adequate dietary intake.

Vitamin C

Also known as ascorbic acid and ascorbate, is a water-soluble antioxidant essential for reducing oxidative stress, supporting immune function, collagen synthesis, and the metabolism of various compounds including amino acids and hormones. Historically, its deficiency disease scurvy caused severe symptoms like impaired wound healing and bleeding gums. Found abundantly in fruits and vegetables, inadequate intake is prevalent, impacting health outcomes such as cardiovascular disease and cancer mortality.

Vitamin D

Vitamin D exists in two major forms: D2 (ergocalciferol) from mushrooms and D3 (cholecalciferol) primarily from animal sources and fortified foods. Synthesized in the skin via UVB radiation and metabolized into its active form, calcitriol, vitamin D plays essential roles in skeletal health, regulating calcium absorption and bone metabolism. Adequacy is assessed by measuring serum 25-hydroxyvitamin D, with optimal levels suggested between 25-80 ng/mL. Deficiency leads to rickets in children and osteomalacia in adults, associated with secondary hyperparathyroidism and bone disorders. Risk factors include insufficient intake, malabsorption, certain medications, and reduced sunlight exposure. Beyond bone health, vitamin D influences immune function, cardiovascular health, insulin production, and may mitigate risks for various chronic diseases.

Vitamin E

Vitamin E, a fat-soluble antioxidant, encompasses multiple tocopherols and tocotrienols, with gamma-tocopherol prevalent in foods. Alpha-tocopherol, widely studied and found in supplements, plays a crucial role in immune function, gene expression, and cardiovascular health, while gamma-tocopherol exhibits anti-inflammatory properties. Found in nuts, seeds, oils, and certain fruits, vitamin E intake in the American diet may be underestimated. Deficiency, rare but linked to fat malabsorption disorders, manifests with neurological symptoms. Clinical trials show dietary vitamin E may aid cardiovascular health, though supplementation's impact remains inconclusive for preventing diseases like Alzheimer's.

Vitamin K

Vitamin K, also known as phylloquinone (K1) and menaquinone (K2), is a fat-soluble vitamin essential for blood clotting and bone metabolism, named from the Danish word "koagulation." Phylloquinone is abundant in dark green leafy vegetables and certain plant oils, while menaquinones, including MK-4 and MK-7, are synthesized in the gut and found in fermented foods like natto, as well as dairy products. Inadequate dietary intake of vitamin K is common in the US, impacting clotting time and potentially bone health. Deficiency may lead to increased prothrombin time and is associated with lower bone mineral density and risks of osteoporosis and fractures.

The 10 Best Vegan Protein Sources

1. Soy? NONGMO

The ubiquitous vegan protein is often associated with processed patties or mystery "meat" loaf, but it doesn't have to be. Soy protein can be a part of a healthy plant-based diet. Foods like tofu, tempeh, edamame and even soy milk are great options for adding protein to your diet. Try cooking up a stir-fry featuring tempeh or tofu, steaming some edamame for an easy appetizer, or topping your morning cereal with soy milk.

According to a 2016 review in Nutrients, many of the health benefits of eating soy are associated with 2-4 servings of soy foods a day. Here is how much protein there is in common soy foods, per the USDA:

- Tempeh: 17 g protein per 1/2 cup
- Shelled edamame: 9 g protein per 1/2 cup
- Tofu: 9 g protein per 3 ounces
- Soy milk: 7 g protein per 1 cup

2. Nutritional Yeast

Don't let nutritional yeast's scientific-sounding name throw you off. Affectionately nicknamed "nooch" by the vegan community, it's an inactive yeast that is yellow in appearance and has a unique cheesy, umami-rich taste. It has 4 grams of protein per 2 tablespoons, according to the USDA, and as a bonus, is a great vegan source of vitamin B12.

Most food sources of vitamin B12 are animal sources, so many vegans need to supplement. Talk to your healthcare provider to make sure you're getting enough if you eat a vegan diet. Enjoy nutritional yeast in sauces or dressings, sprinkled on your next pasta dish or tossed into a bowl of popcorn.

3. Seitan

Seitan is a staple in plant-based diets. It is created with vital wheat gluten, the main protein in wheat, which results in a chewy and hearty texture that really mimics meat in some dishes. It's important to note that because seitan is made with wheat gluten, it is not gluten-free.

A 3-ounce serving of seitan contains 20 grams of protein, per the USDA. You can make seitan yourself by purchasing vital wheat gluten or find it precooked next to the tofu in the refrigerated section of your local supermarket or natural-foods store.

4. Whole Grains

Even though we typically think of them as carbohydrate sources, whole grains can sneak extra protein into any meal. Many varieties are naturally high in protein—not to mention they deliver fiber, vitamins and minerals to your diet. To boost your daily grain intake, start your day with a warm bowl of oatmeal, keep lunch fresh with a quinoa salad or end your evening with wild rice-stuffed peppers for dinner.

Here's a short list of whole grains and how much protein they contain, per the USDA. All measurements are for cooked grains.

- Quinoa: 8 g protein per 1 cup
- Wild rice: 6.5 g protein per 1 cup
- Oats: 6 g protein per 1 cup
- Buckwheat: 5.5 g protein per 1 cup

5. Green Veggies

Often overlooked when it comes to protein, green vegetables offer more than just vitamins and minerals. Foods like spinach, Brussels sprouts and green peas all contain decent amounts of protein to balance out your plate. Not to mention, greens are antioxidant-rich, full of fiber and low in calories. Try adding cooked spinach to pasta, mixing green peas into a curry or roasting up Brussels sprouts for an irresistible crispy side.

Here's a sampling of green veggies and amounts of protein for each, per the USDA. All measurements are for cooked vegetables.

- Spinach: 5 g protein per cup
- Green peas: 4 g protein per 1/2 cup
- Brussels sprouts: 2 g protein per 1/2 cup

6. Sprouted Bread

Sprouted grain bread, also sometimes called Ezekiel bread due to the popular brand name, is a whole-grain baked good that has a hefty amount of protein too. Depending on the brand you purchase, one slice contains 4 to 5 grams of protein, per the USDA. That means that if you make a sandwich with two slices of bread, you're already starting with a whopping 10 grams of protein before you even add the fillings. Other ideas for using sprouted-grain bread include toast, breakfast strata or breadcrumbs.

7. Potatoes

The humble spud isn't always thought of being a health food due to its many unhealthy incarnations (looking at you, french fries and loaded potato skins), but it's actually a wholesome addition to your diet. Just one large russet potato with the skin contains 8 grams of protein, per the USDA that's more potassium than a banana and it's a good source of fiber. Other varieties, like red or sweet potatoes, don't contain as much protein (7 grams and 2.5 grams respectively), but they still can contribute to your daily intake goal. Try potatoes of all types mashed, roasted, baked or scalloped. Here's a recap of protein amounts in potatoes, according to the USDA:

- Russet potato: 8 g per large spud
- Red potato: 7 g per large spud
- Sweet potato: 2.5 g per medium spud

8. Legumes

A go-to for vegans looking to bulk up their protein intake, legumes are the budget-friendly base of many plant-based dishes. The category of legumes includes beans and lentils, both powerhouses when it comes to plant protein. Different lentil varieties can contain up to 18 grams of protein per cup (cooked), while beans can range between 10 and 18 grams per cup depending on the type. Use lentils as taco filling, in chili or as a curry base. Beans are extremely versatile; some of our favorite ways to use them are blended into hummus, formed into fritters or as baked potato toppers.

Here's a brief rundown of lentils and beans and how much protein they contain, per the USDA. All measurements are for cooked legumes.

- Lentils: 18 g per 1 cup
- Chickpeas: 14.5 g per 1 cup
- Black beans: 15 g per 1 cup

9. Seeds

Seeds aren't just for the birds. From sesame seeds whirred into tahini to flax seeds sprinkled onto oatmeal or baked into bread, seeds can be a rich source of protein and fiber in a vegan diet. Flax, chia and hemp are also good sources of plant-based omega-3 fats. Seeds are an especially nice protein option for anyone with nut allergies. Spread sunflower-seed butter on toast, blend tahini into a salad dressing or make a chia seed pudding.

Here are a few seeds and seed butter, including how much protein each contains, per the USDA:

- Pumpkin seeds: 8.5 g per 1 oz
- Hemp seeds: 9.5 g per 3 tablespoons
- Tahini: 5 g per 2 tablespoons

10. Nuts

No plant-based pantry would be complete without several varieties of nuts, which are equally easy to snack on or to incorporate into recipes. The American Heart Association recommends eating 1.5 ounces of nuts or 2 tablespoons of nut butters several times a week. Although the serving sizes are minimal, each contains a hefty dose of protein. Easy uses include packing up pre-portioned baggies of almonds for grab-and-go snacks, whisking peanut butter into sauces and adding a sprinkling of walnuts to your next salad. Here's a sampling of nuts and nut butter and how much protein each contains, per the USDA:

- Almonds: 9 g per 1.5 oz
- Walnuts: 6.6 g per 1.5 oz
- Cashews: 8 g per 1.5 oz
- Peanut butter: 8 g per 2 tablespoons

Vegetable oil - glycerol trifoliate

Vinegar - a dilute solution of acetic acid.

Vulcanization - the process changing the properties of sticky natural rubber into a more usable form with the use of high heat and sulfur.

Wakami Seaweed

Wakame seaweed is rich in vitamins and minerals, including vitamins A, C, D, E, K, and several B vitamins. It also provides minerals such as calcium, magnesium, and iodine. Studies suggest that wakame seaweed has antioxidant properties and may help lower blood pressure and reduce cholesterol levels. It is also known for its potential anti-inflammatory effects.

Walnut

Common tree nut that is a source of protein, monounsaturated fatty acids, polyunsaturated fatty acids, including omega-3-fatty acid, carotenoids, and phytosterols, flavonoids (including proanthocyanidins), phytates, and lignans.

Wasabi

In laboratory studies, Wasabia japonica killed H. pylori, exhibited detoxification, anti-inflammation, and cancer cell apoptosis, including colon cancer cells.

Water

Major component of body fluids whose functions include blood volume and pressure maintenance, body temperature regulation, transport of nutrients, oxygen, waste products, and other substances, chemical reactant, component of secretions, and other roles vital to life and well-being. Hard water is a source of magnesium and calcium, which are associated with maintenance of normal blood pressure, while soft water is higher in sodium and can aggravate high blood pressure and heart disease. Effects of dehydration include metabolic and functional abnormalities ranging from mild dehydration manifesting as nausea and lack of energy to severe dehydration characterized by dark-colored, low-volume urine output, inability to maintain normal body temperature, rapid heart rate, and possible seizure or death. Low water intake is associated with a number of chronic diseases as well.

Watercress

Watercress is a good source of calcium and is a source of lutein and carotenes. Watercress has been traditionally used as an antioxidant, anti-inflammatory, hypolipidemic, and cardioprotective agent. In an experimental study in animals, watercress juice protected cells and did not damage DNA.

Watermelon

Watermelon contains a relatively high sugar content, consisting of sucrose, fructose, and glucose, and is rich in lycopene. A case-control study (n=438 Chinese women age-matched to n=438 controls) that examined dietary intake and breast cancer risk found con- sumption of the "watermelon/papaya/cantaloupe" fruit group was significantly inversely associated with breast cancer risk.

Water Spinach

Water spinach is rich in vitamins A, C, and K, as well as folate and several minerals including iron, calcium, and magnesium. It is also low in calories and carbohydrates.Research indicates that water spinach contains antioxidants that help protect cells from damage. It is also known for its potential to support bone health due to its vitamin K and calcium content.

Wheat germ

Wheat grain embryo that contains high concentrations of unsaturated fats, protein, fiber (3.2 g, 11%), and vitamin E (0.147 mg, 1% DV) per 2 tablespoons. In animal studies, wheat germ improved markers of antioxidant status in animal tissues.

Wheatgrass

Sprouted leaves of wheat that can be homegrown from wheatgrass seed (also called wheatberry) and eaten when the plant is around one or two weeks old. A 1-oz wheatgrass shot is a source of vitamin C (1 mg, 1% DV), vitamin A (94 IU, 3% DV), and phytochemicals such as chlorophyll and flavonoids. Young, newly sprouted seedlings have greater health properties, such as the content of phenolics and flavonoids, compared to older plants. Wheatgrass contains the superoxide dismutase which converts potentially harmful free radical reactive oxygen species into hydrogen peroxides (having extra oxygen molecule to kill cancer cells) and an oxygen molecule. Wheatgrass showed anticancer potential including superoxide scavenging and antioxidative properties in in vitro and animal studies.

Whey

Whey is a significant source of lactose and a good source of calcium, and whey protein is particularly rich in substrates that make the antioxidant glutathione. Whey protein exerted anti-HIV effects in laboratory research; however, in a meta-analysis of various treatments used in HIV-infected children and adults, whey protein concentrate did not significantly alter clinical, anthropometric or immunological outcomes compared with placebo. Whey protein exerted chemoprotective effects and was more effective at reducing body weight gain and increasing insulin sensitivity than red meat in laboratory research.

Wine

Phytochemicals in wine include phenolics, such as resveratrol, quercetin, and catechins, and proanthocyanidins. In Europe, proanthocyanidins are used mainly for the treatment of vascular disorders such as venous insufficiency, varicose veins, and microvascular problems including capillary fragility and retinopathies

Xanthan gum

Xantham gum is a polysaccharide derived from bacteria and is commonly used as a food additive. It is high in fiber and low in calories. Xanthan gum is primarily used as a thickening agent and stabilizer in food products. It has been studied for its potential benefits in improving digestion and managing blood sugar levels in people with diabetes.

Xeaxanthin

Also spelled zeaxanthin. Xanthophyll carotenoid that occurs commonly with lutein in foods. Rich sources include egg yolk, dark green leafy vegetables, orange bell pepper, kiwi fruit, grapes, spinach, zucchini, corn, and squash. Laboratory data suggest inverse correlations between plasma xanthophyll carotenoids and oxidative damage in DNA and lipids.

Xiao Long Bao

Xiao Long Bao is not a significant source of vitamins and is primarily enjoyed for its taste, texture, and cultural significance. While it provides energy and some protein, it is essential to consider its nutritional composition and potential health implications, particularly due to its sodium and fat content. Incorporating a variety of nutrient-rich foods in your diet alongside Xiao Long Bao can help maintain overall nutritional balance and wellbeing.

Yakult

Yakult is a probiotic dairy drink that contains live cultures of the bacterium Lactobacillus casei Shirota. It is low in fat and calories and provides probiotics beneficial for gut health. Yakult has been studied for its potential benefits in supporting digestive health, improving lactose digestion, and boosting the immune system by maintaining a healthy balance of gut bacteria.

Yacon Root

Yacon root is a tuber native to South America and is low in calories, high in fiber, and contains fructooligosaccharides (FOS), which act as prebiotics.Yacon root has been studied for its potential health benefits, including improved digestion, regulation of blood sugar levels, and potential weight loss support due to its prebiotic properties.

Yam

Yams contain folate (34 mcg, 8% DV), are an excellent source of fiber (6.2 g, 22% DV) and vitamin C (26 mg, 28% DV), and a good source of potassium (1,224 mg, 36% DV) per 1 cup. A laboratory study found yam extract to protect against cancer proliferation in human breast cancer cells.

Yogurt

Yogurt is a dairy group food; 1 cup (8 oz) of low-fat vanilla yogurt supplies 388 mg of calcium (29% DV) and no vitamin D (0% DV) because yogurt is not usually fortified with vitamin D.

Yogurt is well tolerated by lactase-deficient subjects resulting in little or no gastrointestinal distress due to "lactase released from the yogurt organisms. The approximately 18 g of lactose in a serving of yogurt produced only 1/3 of the hydrogen excretion in a hydrogen breath test, and subjects consuming it, compared to milk, had fewer reports of diarrhea or flatulence.

Yuca

Also known as cassava, manioc, and tapioca. Starchy tuber that supplies antioxidants, such as betacarotene, resistant starch and is a good source of potassium (271 mg, 5% DV per 100 g), as a worldwide food source of carbohydrate. Catechins and flavone 3-glycosides, suggested to have cardiovascular health benefits, have been identified in cassava.

Za'atar

Za'atar is a Middle Eastern spice blend typically made from dried thyme, oregano, marjoram, sumac, sesame seeds, and salt. It is low in calories and provides various vitamins, minerals, and antioxidants. Za'atar has been studied for its potential antimicrobial, anti-inflammatory, and antioxidant properties, which may contribute to its traditional uses in promoting health and well-being.

Zinc

Trace mineral necessary for immune function, growth, protein and DNA synthesis, wound healing, and taste and sense perception. Doses higher than are in foods may be necessary to restore depleted Zn levels, e.g., due to diarrhea. Features of Zn deficiency are nonspecific and vary according to severity; mild deficiency is characterized by impaired growth as characterized in children by decreased linear growth and inadequate weight gain), appetite, and immune function; additional Zn deficiency symptoms include: diarrhea, hair loss, decreased ability to taste and smell, weight loss, delayed wound healing, and reduced mental alertness.

Zinfandel Grapes

Zinfandel grapes are a variety of black grape used to make red wine. They contain antioxidants like resveratrol and flavonoids, which may have health benefits. Research suggests that moderate consumption of red wine, made from Zinfandel grapes, may be associated with cardiovascular health benefits due to its antioxidant content.

Zucchini

Cooked zucchini supplies 1,000 IU (33% DV) of vita- min A per 1/2 cup, in addition to cucurbitosides, flavonoids, triterpenes, sterols, and beta-cryptoxanthin, one of five predominant carotenoids in plasma.Lutein and xeaxanthin consumption is associated with a decreased risk of age-related macular degeneration. In a laboratory study, betacryptoxan thin exhibited antitumor effects.

ACTN3: MORE THAN JUST A GENE FOR SPEED

Craig Pickering and John Kiely

Abstract

Over the last couple of decades, research has focused on attempting to understand the genetic influence on sports performance. This has led to the identification of a number of candidate genes which may help differentiate between elite and non-elite athletes. One of the most promising genes in that regard is *ACTN3*, which has commonly been referred to as “a gene for speed”. Recent research has examined the influence of this gene on other performance phenotypes, including exercise adaptation, exercise recovery, and sporting injury risk. In this review, we identified 19 studies exploring these phenotypes. Whilst there was large variation in the results of these studies, as well as extremely heterogeneous cohorts, there is overall a tentative consensus that *ACTN3* genotype can impact the phenotypes of interest. In particular, the R allele of a common polymorphism (R577X) is associated with enhanced improvements in strength, protection from eccentric training-induced muscle damage, and sports injury. This illustrates that *ACTN3* is more than just a gene for speed, with potentially wide-ranging influence on muscle function, knowledge of which may aid in the future personalization of exercise training programmes.

Keywords: *ACTN3*, genetics, adaptation, recovery, injury, personalized, genetic testing

Introduction

ACTN3 is a gene that encodes for alpha-actinin-3, a protein expressed only in type-II muscle fibers (North et al., 1999). A common polymorphism in this gene is R577X (rs1815739), where a C-to-T base substitution results in the transformation of an arginine base (R) to a premature stop codon (X). X allele homozygotes are deficient in the alpha-actinin-3 protein, which is associated with a lower fast-twitch fiber percentage (Vincent et al., 2007), but does not result in disease (MacArthur and North, 2004). The XX genotype frequency differs across ethnic groups, with approximately 25% of Asians, 18% of Caucasians, 11% of Ethiopians, 3% of Jamaican and US African Americans, and 1% of Kenyans and Nigerians possessing the XX genotype (Yang et al., 2007; MacArthur et al., 2008; Scott et al., 2010). *ACTN3* genotype is associated with speed and power phenotypes. Yang et al. (2003) reported that elite sprint athletes had significantly higher frequencies of the R allele than controls, a finding that has been replicated multiple times in speed, power and strength athletes (Druzhevskaya et al., 2008; Roth et al., 2008; Eynon et al., 2009; Ahmetov et al., 2011; Cieszczyk et al., 2011; Kikuchi et al., 2016; Papadimitriou et al., 2016; Weyerstraß et al., 2017; Yang et al., 2017), although these findings are not unequivocal (Scott et al., 2010; Gineviciene et al., 2011; Sessa et al., 2011). Whilst Yang et al. (2003) found a trend toward an increased XX genotype frequency in endurance athletes vs. controls, this relationship is less robust, with most studies reporting a lack of association between XX genotype and endurance performance (Lucia et al., 2006; Saunders et al., 2007; Döring et al., 2010; Kikuchi et al., 2016). In addition, whilst Kenyan and Ethiopian endurance runners are highly successful (Wilber and Pitsiladis, 2012), the frequency of the XX genotype within this group is very low at 8% (Ethiopian) and 1% (Kenyan) (Yang et al., 2007). As such, the general consensus is that *ACTN3* X allele likely does not modify elite endurance athlete status (Vancini et al., 2014).

Much of the attention on *ACTN3* has focused on the robust relationship with the R allele and strength/power phenotype, with a number of reviews further exploring this relationship (Eynon et al., 2013; Ma et al., 2013; Ahmetov and Fedotovskaya, 2015). Indeed, a number of papers referenced *ACTN3* as a "gene for speed" (MacArthur and North, 2004; Chan et al., 2008; Berman and North, 2010). However, emerging evidence suggests that this polymorphism may impact a number of other traits, including exercise recovery, injury risk, and training adaptation (Delmonico et al., 2007; Pimenta et al., 2012; Massidda et al., 2017). The purpose of this mini-review is to further explore these potential relationships, as an increased understanding of the role played by *ACTN3* on these traits may lead to improvements in the utilization of genetic information in exercise training.

ACTN3 as a modulator of training response

Over the last 20 or so years, the consistent underlying impact of genetics on exercise adaptation has been well explored (Bouchard et al., 2011; Bouchard, 2012). Whilst it is clear that genetics has an undoubted influence on both exercise performance (Guth and Roth, 2013) and adaptation (Mann et al., 2014), fewer studies examine the influence of individual single nucleotide polymorphisms (SNPs) (Delmonico et al., 2007), or a combination of SNPs (Jones et al., 2016), on this process. In this section, we explore the evidence regarding the impact of *ACTN3* on the post-exercise adaptive response.

Following a structured literature search, we found five studies that examined the influence of *ACTN3* on exercise adaptation to a standardized training programme. Four of these studied resistance training (Clarkson et al., 2005a; Delmonico et al., 2007; Pereira et al., 2013; Erskine et al., 2014), and one focused on aerobic training (Silva et al., 2015). An additional study (Mägi et al., 2016), monitored changes in VO_{2peak} over a five-year period in elite skiers, with no significant *ACTN3* genotype differences. However, the exercise intervention in this study was not controlled, and so we did not include it within. There was considerable variation in the findings. For resistance training, two studies reported that the RR genotype was associated with the greatest increase in strength (Pereira et al., 2013) and power (Delmonico et al., 2007) following resistance training. One study reported no effect of *ACTN3* genotype on training adaptations following resistance training (Erskine et al., 2014). Another reported greater improvement in one-repetition maximum (1RM) in X allele carriers compared to RR genotypes (Clarkson et al., 2005a). A further study utilized *ACTN3* within a 15-SNP total genotype score (TGS), finding that individuals with a higher number of power alleles (such as *ACTN3* R) exhibited greater improvements following high-intensity resistance training compared to low-intensity resistance training (Jones et al., 2016). However, because subjects could have the *ACTN3* XX genotype and still be classed as those who would best respond to high-intensity training (due to the possession of a higher number of alleles in other power-associated SNPs), we did not include this study within.

The variation between studies is likely due to heterogeneity at baseline between genotypes, and differences in exercise prescription. Given the prevalence of the R allele in elite speed-power and strength athletes (Yang et al., 2003; Vincent et al., 2007), it is speculatively considered that R allele carriers would respond best to speed-power and strength training (Kikuchi and Nakazato, 2015). However, as illustrated here, there is perhaps a paucity of data to support this position. Nevertheless, there are some potential molecular mechanisms that could underpin this proposition. Norman et al. (2014) reported that mammalian target of rapamycin (mTOR) and p70S6k phosphorylation was greater in R allele carriers than XX genotypes following sprint exercise. Both mTOR and p70S6k regulate skeletal muscle hypertrophy (Bodine et al., 2001; Song et al., 2005), providing mechanistic support for the belief that hypertrophy, and hence strength and power improvements, should be greater in R allele carriers following resistance training. In addition, Ahmetov et al. (2014) reported that testosterone levels were higher in male and female athletes with at least one R allele compared to XX genotypes. Whilst the direction of this association is not clear, it again supplies a possible mechanism explaining why R allele carriers may experience greater training-induced strength improvements.

A single study examined the impact of this polymorphism on the magnitude of VO_2 improvements following endurance training (Silva et al., 2015). Here, VO_2 scores at baseline were greater in XX genotypes, but following training this difference was eliminated, indicating that RR genotypes had a greater percentage improvement following training. The population in this cohort were police recruits. Given that the X allele is potentially associated with elite endurance athlete status (Yang et al., 2003), it is not clear whether these results would be mirrored in elite endurance athletes. Clearly, further work is required to fully understand what relationship, if any, exists between *ACTN3* and improvements in aerobic capacity following training.

ACTN3 as a modulator of post-exercise recovery

ACTN3 R577X has also been associated with exercise-induced muscle damage; here, increased muscle damage will likely reduce speed of recovery, suggesting a potential modifying effect of this polymorphism on between-session recovery. Of the eight studies identified that examined the impact of this polymorphism on post-exercise muscle damage, six reported that that the X allele and/or the XX genotype was associated with higher levels of markers associated with muscle damage (Vincent et al., 2010; Djarova et al., 2011; Pimenta et al., 2012; Belli et al., 2017; Del Coso et al., 2017a,b). One study found no effect of the polymorphism (Clarkson et al., 2005b), and one found that RR genotypes experienced a greater exercise-induced reduction in force compared to XX genotypes (Venckunas et al., 2012). An additional study (Del Coso et al., 2017c) examined the impact of *ACTN3* as part of a TGS on creatine kinase (CK) response following a marathon race. Within this TGS, the R allele was considered protective against increased CK concentrations. The results indicated that those athletes with a higher TGS, and therefore greater genetic protection, had a lower CK response to the marathon. Whilst not direct evidence of the R allele's protective effect, as it is possible that the other SNPs used in the TGS conveyed this effect, it nevertheless strengthens the supporting argument.

The increase in post-exercise muscle damage is likely due to structural changes associated with this polymorphism. Alpha-actinin-3 is expressed only in fast-twitch muscle fibers, and X allele homozygotes are alpha-actinin-3 deficient; instead, they upregulate production of alpha-actinin-2 in these fast-twitch fibers (MacArthur et al., 2007; Seto et al., 2011). Both alpha-actinin-3 (encoded for by *ACTN3*) and alpha-actinin-2 are major structural components of the Z-disks within muscle fibers (Beggs et al., 1992). The Z-disk itself is vulnerable to injury during eccentric contractions (Friden and Lieber, 2001), and knock-out mouse models illustrates these Z-disks are less stable during contraction with increased alpha-actinin-2 concentrations (Seto et al., 2011). A number of the studies in Table

Table2

2 exclusively utilized eccentric contractions, whilst others focused on prolonged endurance events that include running, which incorporates eccentric contractions as part of the stretch shortening cycle with each stride (Komi, 2000).

The overall consensus of these studies is that the X allele, and/or the XX genotype, is associated with greater markers of muscle damage following exercise that has an eccentric component; either through direct eccentric muscle action (Vincent et al., 2010), from sport-specific training (Pimenta et al., 2012), or from a competitive event requiring eccentric contractions (Belli et al., 2017; Del Coso et al., 2017a,b). However, there are a number of weaknesses to these studies, potentially limiting the strength of these findings. The overall subject number is modest, with a total of 376 (mean 47) across all eight studies; indeed, the study with the greatest number of subjects, Clarkson et al. (2005b), reported no modifying effect of this polymorphism on post-exercise muscle damage. The total number of XX genotypes was also low, with 85 reported across the studies. This is partly a function of the lower prevalence (~18%) of this genotype, but again the study with the largest number (n = 48) of XX genotypes found no effect of this polymorphism (Clarkson et al., 2005b). It is clear that, in order to increase the robustness of this association, further work with greater subject numbers is required.

ACTN3 as a modulator of exercise-associated injury risk

We found six studies examining the association

between *ACTN3* genotype and sports injury risk. Three of these examined ankle sprains (Kim et al., 2014; Shang et al., 2015; Qi et al., 2016), with one each for non-contact injuries (Iwao-Koizumi et al., 2015), professional soccer players (Massidda et al., 2017), and exertional rhabdomyolysis (ER) (Deuster et al., 2013). Whilst ER is strongly related to increased CK following exercise (Clarkson and Ebbeling, 1988; Brancaccio et al., 2010), because it requires medical treatment we classified it as an injury. Of these papers, five reported a protective effect of the R allele and/or the RR genotype against injury (Deuster et al., 2013; Kim et al., 2014; Shang et al., 2015; Qi et al., 2016; Massidda et al., 2017). Specifically, Deuster et al. (2013) found that XX genotypes were almost three times more likely to be ER patients than R allele carriers. Qi et al. (2016) reported a significantly lower frequency of the RR genotype in a group of ankle sprain patients vs. controls. Kim et al. (2014) found that XX genotypes were 4.7 times more likely to suffer an ankle injury than R allele carriers in their cohort of ballerinas. Shang et al. (2015) reported the R allele as significantly under-represented in a cohort of military recruits reporting ankle sprains. Finally, Massidda et al. (2017) demonstrated that XX genotypes were 2.6 times more likely to suffer an injury than RR genotypes, and that these injuries were more likely to be of increased severity. Only one study (Iwao-Koizumi et al., 2015) reported that the R allele was associated with an increased risk (OR = 2.52) of a muscle injury compared to X allele carriers in a female cohort.

Regarding ER, the likely mechanism is similar to that discussed in the post-exercise muscle damage section; increased damage at the Z-disk during exercise. For ankle sprains, the mechanism is potentially related to muscle function. R allele carriers tend to have greater levels of muscle mass (MacArthur and North, 2007), and specifically type-II fibers (Vincent et al., 2007), indicating that both the RX and RR genotypes tend to have increased strength capabilities (Pimenta et al., 2013). For other soft-tissue injury types, again, the decreased potential of damage at the Z-disk likely reduces injury risk. This would be particularly true for eccentric contractions; given the importance of this contraction type in the etiology of hamstring injuries, this could be a further causative mechanism (Askling et al., 2003), alongside that of reduced muscle strength (Yamamoto, 1993).

Alongside the modifying role of *ACTN3* on muscle strength and injury risk, emerging evidence suggests this SNP may also impact flexibility and muscle stiffness. Two studies reported an association between RR genotype and a decreased flexibility score in the sit-and-reach test (Zempo et al., 2016; Kikuchi et al., 2017). Conversely, Kim et al. (2014) reported that XX genotypes had decreased flexibility in the same test. This lack of consensus is largely due to the small total study number, with greater clarity expected as research in the area evolves. It also mirrors the lack of consensus as to whether flexibility increases or decreases risk of injury (Gleim and McHugh, 1997), indicating the complex, multifactorial nature of injuries and their development (Bahr and Holme, 2003).

In summary, it appears that the R allele of *ACTN3* is somewhat protective against injuries. The mechanisms underpinning this are likely varied, and related to a combination of the modifying effects of this SNP on both strength (particularly eccentric strength), exercise-induced muscle damage, and flexibility.

Discussion

The results of this mini-review indicate that, aside from its established role in sporting performance, the *ACTN3* R577X polymorphism also potentially modifies exercise adaption, exercise recovery, and exercise-associated injury risk. As this polymorphism directly influences both muscle structure and muscle fiber phenotype, this is perhaps unsurprising, and points to the potential use of knowledge of this polymorphism in the development of personalized training programmes. However, it is important to consider the limitations surrounding many of these studies. The subject numbers in the considered studies tended to be low, with large heterogeneity between study cohorts, ranging from untrained subjects to professional sports people, as well as differences in sex. Both of these aspects will impact the study findings; the effect of this polymorphism may be smaller in untrained subjects, for example, whereas in elite, well-trained athletes, who are likely closer to their genetic ceiling, the effect may be greater. The low subject numbers are troubling due to the relatively low XX genotype frequency, which is ~18% in Caucasian cohorts, and even lower in African and African-American cohorts. As such, XX genotypes are considerably under-represented across the considered research.

The above limitations indicate further work is required to fully understand the impact of this polymorphism on these phenotypes. That said, there is some consistency between trials, allowing speculative guidelines to be developed for the use of genetic information in the development of personalized training. XX genotypes potentially have increased muscle damage following exercise that includes an eccentric component (Pimenta et al., 2012; Belli et al., 2017; Del Coso et al., 2017a,b). This information may, consequently, be used to guide between-session recovery, and during the competitive season recovery times post-competition. For example, in an elite soccer club, *ACTN3* genotype could be utilized alongside other well-established markers to determine training intensity in the days following a match, with players genetically predisposed to increased muscle damage either having a longer recovery period, or increased recovery interventions such as cold-water immersion. In addition, recent research has illustrated the positive impact of Nordic Hamstring Exercises on hamstring injury risk (van der Horst et al., 2015), making these exercises increasingly common in professional sports teams. These exercises have a large eccentric component, upon which this polymorphism may have a direct effect. As such, it would be expected that XX genotypes would have increased muscle soreness and damage following these exercises, potentially impacting the timing of their use within a training programme.

Focusing on sporting injuries, the general consensus from the studies found is that the X allele increased the risk of ankle injuries (Kim et al., 2014; Shang et al., 2015; Qi et al., 2016) and general sporting injury (Massidda et al., 2017). Again, this information could guide training interventions. In this case, X allele carriers might undertake increased general strengthening exercises and neuromuscular training targeting injury risk reduction. Furthermore, knowledge of this information could increase athlete motivation to undertake these exercises (Goodlin et al., 2015).

Finally, maximizing the training response is crucial, both to elite athletes looking to improve by fractions of a second, and to beginners looking to decrease their risk of disease. Increasingly, there is evidence that polymorphisms, including *ACTN3* R577X, can impact this adaptive process (Delmonico et al., 2007; Pereira et al., 2013). If further research replicates these early findings, then again, this information could be used in the development of training programmes. Regarding *ACTN3*, at present it appears that R allele carriers potentially exhibit greater increases in strength and power following high-load resistance training (Delmonico et al., 2007). As such, Kikuchi and Nakazato (2015) speculate that R allele carriers should prioritize high-load, low-repetition resistance training if improvements in muscle strength are required, and high intensity interval (HIT) training to specifically elicit improvements in VO_{2max}.

Conclusion

There is a clear, undoubted impact of genetics on both sporting performance and exercise adaptation. In this regard, one of the most well-studied genes is *ACTN3*, which has been reliably shown to impact speed-power and strength phenotypes. However, emerging research indicates that this polymorphism may also impact other exercise associated variables, including training adaptation, post-exercise recovery, and exercise-associated injuries.

This information is important, not just because it illustrates the wide-ranging impact SNPs can have, but also because it represents an opportunity to personalize, and therefore enhance, training guidelines. At present, there are no best-practice guidelines pertaining to the use of genetic information in both elite sport and the general public. However, sports teams have been using genetic information for over 10 years (Dennis, 2005), and continue to do so. Consequently, the development of these guidelines represents an important step from lab to practice. Clearly, further research is required to fully develop these guidelines, and at present such information is speculative. Nevertheless, the use of genetic information represents an opportunity to enhance training prescription and outcomes in exercisers of all abilities.

THE NEW GENETICS OF INTELLIGENCE

Robert Plomin and Sophie von Stumm

Abstract

Intelligence — the ability to learn, reason and solve problems — is at the forefront of behavioural genetic research. Intelligence is highly heritable and predicts important educational, occupational and health outcomes better than any other trait. Recent genome-wide association studies have successfully identified inherited genome sequence differences that account for 20% of the 50% heritability of intelligence. These findings open new avenues for research into the causes and consequences of intelligence using genome-wide polygenic scores that aggregate the effects of thousands of genetic variants. In this Review, we highlight the latest innovations and insights from the genetics of intelligence and their applications and implications for science and society.

Introduction

Life is an intelligence test. During the school years, differences in intelligence are largely the reason why some children master the curriculum more readily than other children. Differences in school performance predominantly inform prospects for further education, which in turn lead to social and economic opportunities such as occupation and income. In the world of work, intelligence matters beyond educational attainment because it involves the ability to adapt to novel challenges and tasks that describe the different levels of complexity of occupations. Intelligence also spills over into many aspects of everyday life such as the selection of romantic partners and choices about health care[1]. This is why intelligence — often called general cognitive ability[2] — predicts educational outcomes[3], occupational outcomes[4,5] and health outcomes[6] better than any other trait. It is also the most stable psychological trait, with a Pearson correlation coefficient of 0.54 from age 11 years to age 90 years[7].

What is intelligence?

Intelligence can be broadly defined as the ability to learn, reason and solve problems[74]. It is a latent trait that cannot be directly observed but is inferred from a battery of diverse cognitive test scores, as in widely used 'intelligence tests 'that yield a so-called 'IQ score', which is an acronym for an outdated concept of an 'intelligence quotient'. Psychometric tests of cognitive abilities differ widely in form and content. For example, some assess verbal ability and others non-verbal ability, some give strict time limits and some are untimed (see figure for examples). Notwithstanding these differences, cognitive test scores are positively inter-correlated[75], suggesting that any differences in test scores that occur within an individual are smaller than test score differences that exist between individuals. In other words, a person who scores high on one type of cognitive test relative to other people will also do comparatively well on other cognitive tests. This phenomenon is known as the positive manifold, or simply *g*, the general factor of intelligence, which emerges from the test scores' covariance, discovered by Spearman in 1904[76], about the same time that Mendel's laws of inheritance were rediscovered. The g-factor exemplifies the generalist nature of intelligence as a complex trait that penetrates many behavioural and psychological outcomes, including educational attainment, occupational status, health and longevity[77,78].

Individual differences in intelligence are fairly stable across the lifespan, especially from teenage years onwards, with correlations of 0.6 and above[38,79]. However, intelligence is also subject to change, both within and between individuals. For example, scores from timed cognitive tests tend to peak in young adulthood and decline thereafter[80]. But more importantly, intelligence has been shown to be malleable, especially in children, through major systematic interventions, such as education[81], dietary supplementation[82] or adopting children away from impoverished home environments[83]. That said, identifying ways to effectively improve intelligence remains a key challenge for intelligence research, with many interventions failing to produce reliable and long-term positive effects[82,84,85].

During the past century, genetic research on intelligence was in the eye of the storm of the nature–nurture debate in the social sciences[8,9]. In the 1970s and 1980s, intelligence research and its advocates were vilified[10–12]. The controversy was helpful in that it raised the quality and quantity threshold for the acceptance of genetic research on intelligence. As a result, bigger and better family studies, twin studies, and adoption studies amassed a mountain of evidence that consistently showed substantial genetic influence on individual differences in intelligence[13]. Meta-analyses of this evidence indicate that inherited differences in DNA sequence account for about half of the variance on measures of intelligence[14].

These studies and applications in neuroscience[15] were already pushing intelligence research towards rehabilitation when it was thrust to the forefront of the DNA revolution 4 years ago by genome-wide association studies (GWAS) focused on a very different variable; years of education. In this Review, we discuss early attempts to find the inherited DNA differences that account for the substantial heritability of intelligence, and a twist of fate involving GWAS on years of education, before discussing the results of recent large GWAS of intelligence. The second half of this Review focuses on genome-wide polygenic scores (GPS) for intelligence that aggregate the effects of thousands of DNA variants associated with intelligence across the genome (see Box 2 for how GPS are constructed). We illustrate how GPS for intelligence will transform research on the causes and consequences of individual differences in intelligence, before ending with a discussion of societal and ethical implications. We do not discuss other important research related to intelligence such as evolutionary research[16,17] and neuroscience research[15] in order to focus on the role of GPS in the new genetics of intelligence.

Creating GPS

Thousands of SNPs are needed to account for the

heritability of intelligence and other complex traits because the effect sizes of SNP associations are so small. Aggregating thousands of these miniscule effects in a GPS is the crux of the new genetics of intelligence. There are at least a dozen labels to denote GPS. Most involve the word 'risk', such as polygenic risk scores. We prefer the term genome-wide polygenic score (GPS) because 'risk' does not apply to quantitative traits, such as intelligence, that have positive as well as negative poles86. The 'genome-wide 'addition to 'polygenic score 'distinguishes GPS from polygenic scores that aggregate candidate genes or just the top hits from GWAS. Finally, another reason for using the acronym GPS is that we cannot resist the metaphor of the other 'GPS', global positioning system. We see IQ GPS as a system to triangulate on the genetics of intelligence from all domains of the life sciences.

An intelligence test score is a composite of several tests, often with each test weighted by its contribution to general intelligence. In the same way, a GPS is a composite of SNP associations, weighted by their correlation with the trait. The table shows how a GPS could be constructed for one individual for 10 SNPs. GWAS results are used to determine which of the two alleles for a SNP is positively associated with the trait, called the increasing allele. For each SNP, a genotypic score is created by adding the number of increasing alleles. A GPS sums the number of increasing alleles across SNPs — hence, why this is called an additive model. In this example, the individual's GPS is 9. Because there are 10 SNPs, the possible range of the GPS is from 0 to 20.

A more predictive GPS can be constructed by weighting

each genotypic score by the effect size of the SNP (beta for quantitative traits, odds ratio for qualitative traits) as gleaned from GWAS results (see table). For instance, for SNP 1, the correlation with the trait is five times greater than for SNP 10. Multiplying the genotypic score by the correlation gives a weighted genotypic score (see table, last column). Summing these weighted genotypic scores gives this individual a GPS of 0.023 for intelligence. Other ways to improve the predictive power of GPS include taking into account expected SNP effect sizes, the genetic architecture of the trait and specifically modelling linkage disequilibrium[87]. Programmes including LDpred[88] and PRSice[89] provide pipelines for the construction of GPS.

How many SNPs should be included in a GPS? The goal is to maximize predictive power in samples independent from the GWAS samples. Using only genome-wide significant hits does not predict nearly as well as using tens of thousands of SNPs. LDpred uses all SNPs, imputed as well as measured, although most SNPs are given near-zero weights.

Once a GPS for intelligence is created for each individual in a sample, it can be used like any other variable in analyses. For example, it can be used to investigate the extent to which this genetic index of intelligence mediates or moderates effects on variables of primary interest to the researcher.

Finding the heritability of intelligence

Similar to many other complex traits, early results for intelligence were disappointing for more than 100 candidate gene studies[18] and for seven GWAS[19–25]. From the 1990s until 2017 no replicable associations were found. GPS from these early GWAS, which we refer to as 'IQ1', predicted only 1% of the variance of intelligence in independent samples. It became clear that the problem was power: the largest effect sizes of associations between individual single nucleotide polymorphisms (SNPs) and intelligence were extremely small, accounting for less than 0.05% of the variance of intelligence. The average effect size of the tens of thousands of SNPs needed to explain the 50% heritability of intelligence is of course much lower. If the average effect size is 0.005%, 10,000 such SNP associations would be needed to explain the 50% heritability of intelligence. To achieve sufficient power for detecting such tiny effect sizes (that is, power of 80%, P = 0.05 one-tailed), sample sizes greater than 250,000 are required. IQ1 GWAS had sample sizes from 18,000 to 54,000, which seemed large at the time but were not sufficiently powered to detect such small effects.

Breakthrough for years of education

A breakthrough for intelligence research came from the unlikely variable of the number of years spent in full-time education, often referred to as educational attainment. Because 'years of education 'is obtained as a demographic marker in nearly every GWAS, it was possible to accumulate sample sizes with the necessary power to detect very small effect sizes[26]. Its relevance to intelligence is that years of education is highly correlated phenotypically (0.50) and genetically (0.65) with intelligence[27].

In 2013, a meta-analytic GWAS analysis of years of

education yielded three genome-wide significant SNP associations in a sample of 125,000 individuals from 54 cohorts[28]. These associations could be replicated in independent samples[29]. The largest effect size associated with an individual SNP accounted for a meagre 0.02% of the variation, equivalent to about 2 months of education. Although individual SNPs of such miniscule effect size are fairly useless for prediction, a GPS based on all SNPs regardless of the strength of their association with years of education predicted 2% of the variance in years of education in independent samples[28,29]. We refer to this GPS as EA1.

Spurred on by this success, in 2016, a second meta-analytic GWAS analysis with a sample size of 294,000 identified 74 significant loci[30]. This analysis produced a GPS, EA2, that predicted 3% of the variance in years of education on average in independent samples[30]. Surprisingly, GPS for years of education predicted more variance in intelligence than they predicted for the GWAS target trait of years of education[27]. For example, EA2 GPS predicts 3% of the variance in years of education but it predicts 4% of the variance in intelligence[30]. A third GWAS currently in progress includes more than one million participants, making it the largest GWAS for any trait to date. Preliminary results from this GWAS have identified more than 1,000 significant associations and a GPS, EA3, that predicts more than 10% of the variance in years of education in independent samples (Philipp D. Koellinger, personal communication)[31]. Hence, EA3 GPS is expected to predict more than 10% of the variance in intelligence. The effect size of EA3 GPS for predicting intelligence is likely to rival that of family socioeconomic status, which is indexed by parents' years of education. Across studies, parents' education correlates 0.30 with children's intelligence, implying that it accounts for 9% of the variance in children's intelligence[38]. This association is, however, confounded by genetics, because children inherit the DNA differences that predict their intelligence from their parents. Furthermore, parental phenotypes, such as education, only estimate an average association for offspring, whereas GPS predict intelligence for each individual.

Large-scale GWAS of intelligence

In 2017, the largest GWAS meta-analysis of intelligence, which included 'only '78,000 individuals, yielded 18 genome-wide significant regions[32]. A GPS (IQ2) derived from these GWAS results finally broke the 1% barrier of previous GWAS of intelligence by predicting 3% of the variance of intelligence in independent samples. However, IQ2 still has less predictive power than the 4% of the variance explained by the EA2 GPS.

A follow-up GWAS meta-analysis reached a sample size of 280,000 with the inclusion of cognitive data from the UK Biobank (www.ukbiobank.ac.uk). This GWAS analysis increased the number of identified genome-wide significant regions from 18 to 206[33]. A GPS derived from these GWAS analyses, IQ3, predicts about 4% of the variance of intelligence in independent samples[33]. Other meta-analytic GWAS using the UK Biobank data, which were released in June 2017 and are publicly available, yield similar results[34].

These IQ and EA GPS results are summarized in Figure

1. It might seem disappointing that the increase of the intelligence GWAS sample sizes from 78,000 to 280,000 only boosted the predictive power of the IQ GPS from 3% to 4%. However, this result is parallel to GWAS results for years of education: after increasing sample sizes from 125,000 to 294,000, the variance in years of education predicted by the EA GPS only grew from 2% to 3%. Note that the predictive power of EA GPS jumped to more than 10% of the variance in preliminary analyses of the latest meta-analytic GWAS (EA3) with a sample size of over one million (Philipp D. Koellinger, personal communication). We can expect a similar jump in predictive power of the IQ GPS when the sample size for GWAS meta-analyses of intelligence exceeds one million. However, it is more difficult to obtain huge sample sizes for intelligence, which has to be tested, than for years of education, which can be assessed with a single self-reported item.

Missing heritability

It is possible to use multiple GPS to boost the power to predict intelligence by aggregating GPS in a way analogous to aggregating SNPs to produce GPS (Box 3). Including EA2 GPS, IQ2 GPS and other GPS in this multivariate way can already predict up to 7% of the variance in intelligence[35,36]. Multivariate GPS analyses that incorporate multiple GPS in addition to EA2 GPS and IQ2 GPS will explain substantially more than 10% of the variance in intelligence, which is more than 20% of the 50% heritability of intelligence.

Using multiple GPS to predict a trait

Aggregating thousands of SNP associations in GPS has been key to predicting individual differences in complex traits such as intelligence. In an analogous manner, it is possible to aggregate many GPS to exploit their joint predictive power. For example, multiple GPS were used to predict intelligence in a sample of 6,710 unrelated 12-year-olds[35]. This approach is a multiple regression prediction model that accommodates multiple correlated predictors while preventing overfitting based on training in one sample and testing in another sample in a repeated cross-validation design. This approach predicted 4.8% of the variance in intelligence. Although EA2 GPS alone accounted for most of the variance, other GPS added significantly to the prediction of intelligence, especially GPS derived from GWAS of high-IQ individuals [57], childhood IQ[19] and household income[90]. More than 7% of the variance in intelligence was predicted[36] using another approach called Multi-Trait Analysis of GWAS (MTAG)[91], which performs a meta-analysis from summary statistics for a few correlated GPS and produces new summary statistics that can be used to create a multivariate GPS.

The success of GWAS came from its atheoretical approach that analyses all SNPs in the genome rather than selecting candidate genes. In the same way, an atheoretical approach can be used in analyses of multiple GPS by incorporating as many GPS as possible rather than selecting a few candidate GPS. For example, the first study of this sort, mentioned above[35], included a total of 81 GPS from well-powered GWAS of cognitive, medical and anthropometric traits available in LD Hub[92] that together predicted 4.8% of the variance in intelligence. Although EA2, IQ and income GPS drove most of the predictive power of this multiple-GPS analysis, significant independent contributions to the prediction of intelligence were also found for major depressive disorder GPS and autism spectrum disorder GPS. These latter associations were in the direction expected based on the negative genetic correlation between intelligence and depression and the surprising positive genetic correlation between intelligence and autism (see Multivariate genetic research).

Nonetheless, 10% is a long way from the heritability estimate of 50% obtained from twin studies of intelligence[14]. This lacuna is known as the 'missing heritability', which is a key genetic issue for all complex traits in the life sciences[39] (Box 4). The current limit for the variance that can be predicted by GPS is SNP heritability, which estimates the extent to which phenotypic variance for a trait can be explained by SNPs across the genome without identifying specific SNP associations. For intelligence, SNP heritability is about 25%[33,40]. It is safe to assume that GPS for intelligence using current SNP chips can approach the SNP heritability limit of 25% by amassing ever-larger GWAS samples and by using multi-trait GWAS that include traits related to intelligence such as years of education. However, breaking through this ceiling of 25% SNP heritability to the 50% heritability estimated from twin studies — assuming that twin studies yield accurate estimates of the total variance explained by inherited DNA differences — will require different technologies, such as whole-genome sequencing data that include rare variants, not just the common SNPs used on current SNP chips.

Twin, SNP and GPS heritabilities

Heritability is the proportion of observed (phenotypic) differences among individuals that can be attributed to genetic differences in a particular population. Broad heritability involves all additive and nonadditive sources of genetic variance, whereas narrow heritability is limited to additive genetic variance. Additive genetic variance refers to the independent effects of alleles or loci that 'add up'. Nonadditive genetic variance involves effects of alleles or loci that interact.

Twin heritability compares the resemblance of identical and fraternal twins to estimate genetic and environmental components of variance. For intelligence, twin estimates of broad heritability are 50% on average[14]. Adoption studies of first-degree relatives yield similar estimates of narrow heritability of intelligence, suggesting that most genetic influence on intelligence is additive.

SNP heritability is estimated directly from SNP differences between individuals. It does not specify which SNPs are associated with a trait. Instead, it uses chance genomic similarities across hundreds of thousands of SNPs genotyped on a SNP chip for thousands of unrelated individuals to estimate the extent to which genomic covariance accounts for phenotypic covariance in these individuals. For intelligence, SNP heritability is about 25%[22,33,40].

GPS heritability is the proportion of variance that can be predicted by the GPS. For intelligence, GPS heritability is currently about 10% (P. D. Koellinger, personal communication).

These three types of heritability denote two types

of 'missing heritability, 'as shown in the figure. SNP heritability is the ceiling for GWAS and for GPS heritability because all three rely on the additive effects of SNPs genotyped on SNP chips[93]. The missing heritability gap between GPS heritability (10%) and SNP heritability (25%) can be narrowed by increasing GWAS sample size. Narrowing the missing heritability gap between SNP heritability (25%) and twin heritability (50%) will require different technologies that consider, for example, rare variants, gene-gene interaction, and gene-environment interaction.

GPS in intelligence research

The extremely polygenic nature of intelligence means that it will be a long slog from genome-wide significant 'hits' found across GWAS of intelligence. A bottom-up approach focused on specific genes will be difficult for two reasons. First, many hits are in intergenic regions, which means that there are no 'genes 'to trace through the brain to behaviour. Second, the biggest hits have miniscule effects — less than 0.05% of the variance — which means that hundreds of thousands of SNP associations are needed to account for the 50% heritability estimated by twin studies. A systems biology approach to molecular studies of the brain is needed that is compatible with this extreme polygenicity[41].

By contrast, the top-down approach of GPS that aggregate thousands of these tiny effects is already transforming research on intelligence[42]. Unlike quantitative genetic studies that require special samples such as twins, or GWAS that require huge samples in the hundreds of thousands, GPS can be used to add a genetic dimension to any research with modest sample size. For example, a GPS for intelligence that predicts 10% of the variance only needs a sample size of 60 to detect its effect with 80% power (P =.05, one-tailed).

GPS are unique predictors in the behavioural sciences. They are an exception to the rule that correlations do not imply causation in the sense that there can be no backward causation when GPS are correlated with traits. That is, nothing in our brains, behaviour or environment changes inherited differences in DNA sequence. A related advantage of GPS as predictors is that they are exceptionally stable throughout the life span because they index inherited differences in DNA sequence. Although mutations can accrue in the cells used to obtain DNA, like any cells in the body these mutations would not be expected to change systematically the thousands of inherited SNPs that contribute to a GPS.

In other words, a GPS derived from a GWAS of any trait at any age would be expected to correlate near 1.0 when the GPS is constructed from DNA obtained at birth and in adulthood for the same individual, although we are not aware of any empirical evidence relevant to this prediction. If GPS for individuals do not change during the life span, a GPS derived from GWAS of intelligence in adulthood will predict adult intelligence just as well from DNA obtained at conception or birth as from DNA obtained in adulthood. By contrast, intelligence tests at birth cannot predict intelligence at age 18 years. At 2 years of age, infant intelligence tests predict less than 5% of the variance of intelligence in late adolescence[37,38].

GPS are unbiased in the sense that they are not subject to training, faking or anxiety. They are also inexpensive, costing less than US$100 per person. This expense would not be incurred specifically to predict intelligence; the same SNP chip genotype information used in GWAS can be used to create GPS for hundreds of disorders and traits, one of which is intelligence.

GPS for intelligence will open new avenues for research into the causes and consequences of intelligence. Three examples are developmental change and continuity, multivariate links between traits, and gene–environment interplay. A critical requirement for capitalizing on these opportunities is to make the ingredients for GPS publicly available — that is, GWAS summary-level statistics.

Make GWAS summary-level statistics publicly available
It is essential for continued rapid scientific advances using GPS that summary-level statistics from GWAS are made publicly available for all SNPs following publication. The reason why public access to summary statistics is important is that the construction of GPS requires an effect size indicator and *P* value for each SNP in the GWAS. GWAS summary-level statistics are also necessary for other analyses, most notably LD score regression, which is used to estimate genetic correlations among traits[54].

Until 2017, GWAS summary-level data were stored in different databases using different formats, which made it difficult to use the data to investigate traits across studies. This problem has been solved with LD Hub, a centralized database and web interface that provides an automated pipeline for entering and using GWAS summary-level data[92].

However, only about 10% of published GWAS results are publicly available on LD Hub. Some GWAS consortia are exemplars for making GWAS summary-level data immediately upon publication, or even before publication, such as the Psychiatric Genomics Consortium[94]. In intelligence research, a paragon is the Social Science Genetic Association Consortium[27], which is responsible for five of the six GWAS for which summary statistics are publicly available in the intelligence section of LD Hub, although three of the five GWAS were for years of education rather than for intelligence itself.

In contrast, some authors apply conditions for the use of the summary statistics from their published GWAS paper. Others refuse to share these statistics altogether. A worrying trend is that several commercial organizations do not allow summary GWAS statistics from their samples to be used in open-access summary-level statistics for all SNPs when their samples are included in meta-analytic GWAS. Concerns about privacy have been put forth as an explanation, but these fears should be allayed as it is not possible to re-construct individual-level data from summary-level GWAS statistics in large heterogenous samples[95].

Such asymmetrical data-sharing policies between industry and academia will hold back research in the field. If a group does not want their summary-level GWAS statistics to be freely available for a published meta-analytic GWAS, their data should not be used in 'publication', true to its Latin origin publicare, which means 'to make public'.

Developmental research

One of the most interesting developmental findings about intelligence is that its heritability as estimated in twin studies increases dramatically from infancy (20%) to childhood (40%) to adulthood (60%), while age-to-age genetic correlations are consistently high[43,44]. What could account for this increasing heritability despite unchanging age-to-age genetic correlations? Twin studies suggest that genetic effects are amplified through gene–environment correlation as time goes by[45]. That is, the same large set of DNA variants affects intelligence from childhood to adulthood, resulting in high age-to-age genetic correlations, but these DNA variants increasingly have an impact on intelligence as individuals select environments correlated with their genetic propensities, leading to greater heritability of intelligence.

Developmental hypotheses about high age-to-age genetic correlations and increasing heritability can be tested more rigorously and can be extended using GPS. Does the variance explained by GPS for intelligence increase from childhood to adolescence to adulthood? Are the correlations between GPS at these ages consistently high?

High age-to-age genetic correlations for intelligence imply that GWAS of adults should predict intelligence in childhood. The EA2 GPS[30], currently the best genetic predictor of intelligence until EA3 GPS becomes available, was derived from a GWAS meta-analysis of years of education in adults who had completed their education. Nonetheless, the EA2 GPS predicts 2% of the variance in intelligence at age 7 years, 3% at age 12 years, and 4% at age 16 years in our longitudinal study[46].

Multivariate genetic research

Multivariate genetic research focuses on the genetic covariance between traits rather than the variance of each trait. A specific multivariate question for intelligence research is why EA GPS predict twice as much variance in intelligence as do GPS for intelligence itself. This question raises interesting methodological and conceptual issues.

EA GPS and intelligence

EA GPS predict intelligence because the genetic correlation between years of education and intelligence is greater than 0.50 in twin studies[97] and LD score regression studies[98]. The genetic correlation of 0.50 also sets a limit on the extent to which EA GPS can predict intelligence.

But why do EA GPS predict intelligence to a greater extent than they predict EA itself? That is, EA2 GPS predicts 3% of the variance in years of education [30] but it predicts 4% of the variance in intelligence[46]. Moreover, EA GPS predict intelligence much better than IQ GPS predict intelligence themselves. The IQ3 GPS from the most recent GWAS of intelligence predicts 4% of the variance of intelligence[33] but the EA3 GPS predicts more than 10% of the variance in intelligence (P. D. Koellinger, personal communication).

There are two likely reasons why EA GPS currently predict intelligence to a greater extent than EA GPS predict years of education itself. First, intelligence may be more heritable (60% in adults) than years of education (40%) in twin studies[99]. Second, years of education is a coarse measure, primarily indicating whether an individual completed university. Years of education is largely bimodal, with a spike at the end of secondary school and another peak for individuals who attended university. By contrast, intelligence is a more refined measure than years of education that captures the commonalities among diverse tests of cognitive abilities and is normally distributed. That is, educational achievement is not just a proxy for intelligence. It is also predicted by personality traits such as conscientiousness and well-being and having fewer mental health problems such as depression. Together, these non-cognitive traits account for as much of the heritability of educational achievement as intelligence[96]. The EA GWAS incorporates SNPs associated with any of these traits, not just with intelligence[14].

Of note, the current GWAS sample sizes for EA are three times larger than for intelligence. The GPS effect sizes for intelligence are similar to those for EA GPS for comparable effect sizes (that is, IQ2 as compared to EA1, and IQ3 as compared to EA2; see Fig. 1). For this reason, we predict that an IQ GPS derived from a GWAS of intelligence with a sample size of one million, such as EA3, will predict at least as much variance in intelligence as does the current EA3 GPS. In other words, intelligence is not actually predicted to a greater extent by EA GPS than by intelligence GPS when the powers of the discovery GWAS are similar.
Multivariate genetic research is especially important for intelligence because genetic effects in the cognitive domain have been shown in twin studies to be general. That is, genetic effects correlate highly across most cognitive abilities such as verbal and spatial abilities as well as most educational skills such as reading and

mathematics[47]. A recent multivariate finding is that the EA2 GPS predicts 5% of the variance in comprehension and efficiency of reading[48]. This is by far the most powerful GPS predictor of reading ability because there have as yet been no large GWAS of reading with replicable results[49]. EA GPS are also likely to predict other educational skills such as mathematics and other cognitive abilities such as spatial ability.

EA GPS do not only predict reading. They are correlated genetically with a wider range of variables than any other trait[50]. This pervasive genetic influence of EA GPS extends to a negative genetic correlation with schizophrenia and positive genetic correlations with height[51], myopia[52], and surprisingly with autism[53]. Linkage disequilibrium (LD) score regression analysis[54], which uses summary GWAS statistics rather than GPS for individuals, finds a similar pattern of results for intelligence using the IQ2 GWAS: the negative genetic correlation with schizophrenia (–0.20) and the positive genetic correlations with height (0.10) and autism (0.21)[32]. The same LD score regression analysis[32] found that intelligence significantly correlated genetically with many other traits, including Alzheimer disease (–0.36), smoking cessation (–0.32), intracranial volume (0.29), head circumference in infancy (0.28), depressive symptoms (–0.27), attention-deficit-hyperactivity disorder (–0.27), ever smoked (–0.23), longevity (0.22) and, of course, years of education (0.70).

Despite this evidence for ability-general genetic effects, genetic correlations across cognitive abilities and educational skills are not 1.0, which implies that there are ability-specific SNP associations. An important direction for research is to identify ability-specific GPS derived from large GWAS analyses focused on specific cognitive abilities independent of general intelligence. Preliminary analyses of this sort would be possible using existing GWAS of intelligence because most of these studies assessed multiple measures of specific cognitive abilities, which were combined to index intelligence. These data could be re-analysed in meta-analytic GWAS that focus on specific abilities included in multiple studies. However, what is needed are large GWAS focused on well measured specific cognitive abilities such as verbal, spatial and memory abilities and specific cognitive skills taught in schools such as reading, mathematics and language. The pay-off from these studies will be GPS that predict specific abilities independent of general intelligence. These ability-specific GPS could be used to create profiles of genetic strengths and weaknesses for individuals that could be the target for personalized prediction, prevention and intervention.

In addition to investigating links between different traits, multivariate genetic research can examine genetic links between dimensional and diagnostic measures of the 'same 'domain. For example, EA2 GPS predicts reading disability just as much as reading ability, from slow readers to speed-readers[48]. Because GPS are always normally distributed, they will show that there are no etiologically distinct common disorders, only continuous dimensions[55]. This is also true for very low and for very high intelligence[46]. Even extremely high intelligence is only quantitatively, not qualitatively, different genetically from the normal distribution[56,57]. The exception is severe intellectual disability, which is genetically distinct from the rest of the distribution of intelligence[58] and affected by rare, often de novo, mutations of relatively large effect[59].

Research on gene–environment interplay

The high heritability of intelligence should not obscure the fact that heritability is significantly less than 100%. Research using genetically sensitive designs has led to one of the most important findings about environmental influence on intelligence. Intelligence has always been known to run in families but it was assumed that this family resemblance was due to nurture, called shared family environmental influence. That is, siblings were thought to be similar in intelligence because they grew up in the same family and attended the same schools. Twin and adoption studies consistently support this assumption but only up until adolescence. After adolescence, the effect of shared family environmental influence on intelligence is negligible, which means that family environments have little effect on individual differences in the long run[45,60]. Family resemblance for intelligence is due to nature rather than nurture, although it should be emphasized that we are referring to the normal range of environmental influence, not the extremes such as neglect or abuse. However, little is known about the specific environmental factors that make children growing up in the same family different[14].

The importance of both genetics and environment for cognitive development recommends investigating the interplay between them. GPS for intelligence will greatly facilitate this research because they offer, for the first time, the possibility of directly assessing genetic propensities of individuals to investigate their interplay with aspects of the environment. Gene–environment (GE) interplay refers to two different concepts, GE interaction and GE correlation.

GE interaction denotes a conditional relationship in which the effects of genes on intelligence depend on the environment. For example, some twin research suggests that heritability of intelligence is lower in low socioeconomic status family environments and higher in high socioeconomic status family environments[61]. This hypothesis predicts that GPS for intelligence will correlate less with intelligence in environments of low socioeconomic status compared to those with high socioeconomic status. The first test of this hypothesis using the EA2 GPS found no evidence for such an interaction[46]. That is, EA2 GPS were just as much correlated with intelligence in low socioeconomic status as in high socioeconomic status family environments. GPS provide a particularly powerful approach to test for GE interaction as compared to twin studies[62].

In contrast to GE interaction, GE correlation refers to the correlation between genetic propensities and experiences. GE correlation is the reason why most environmental measures used in the behavioural sciences show genetic influence in twin studies[63]. Associations between environmental measures and behavioural traits such as intelligence are also mediated in part by genetic differences. Research using GPS is beginning to confirm these twin study findings about the 'nature of nurture 'by showing, for example, that EA GPS correlate with social mobility[64] and capture covariation between environmental exposures and children's behaviour problems and educational achievement[65]. GE correlation provides a general model for how genotypes become phenotypes — how children select, modify and create environments correlated with their genetic propensities. GPS will greatly advance research on GE correlation by providing an individual-specific index of the 'G' of GE interplay. GPS will also make it possible to assess environmental influences on intelligence while controlling for genetic influences.

Implications for society

The most exciting aspect of GPS is their potential for addressing novel, socially important questions, which we will illustrate with three recent examples from our own research. First, children in public and private schools differ in their EA2 GPS scores because private schools select pupils based on genetic differences in intelligence[66]. Second, intergenerational educational mobility reflects EA2 GPS differences[67]. Finally, the EA2 GPS predicts twice as much variance in educational attainment and occupational status in the post-Soviet era as compared to the Soviet era in Estonia, a finding compatible with the hypothesis that heritability is an index of equality of opportunity and meritocracy[68].

Understanding ourselves

IQ GPS will be used to predict individuals 'genetic propensity to learn, reason and solve problems, not only in research, but also in society, as direct-to-consumer genomic services provide GPS information that goes beyond single-gene and ancestry information. We predict that IQ GPS will become routinely available from direct-to-consumer companies along with hundreds of other medical and psychological GPS that can be extracted from genome-wide genotyping on SNP chips. Using GPS to predict individuals 'genetic propensities requires clear warnings about the probabilistic nature of these predictions and the limitations of their effect sizes.

GPS must be used with caution when predicting outcomes in individuals. We illustrate the probabilistic nature of GPS predictions using data on EA2 GPS and school achievement from the Twins Early Development Study[100]. School achievement was assessed by scores from a UK-wide examination, the General Certificate of Secondary Education (GCSE), administered at the end of compulsory education at age 16 years. GCSE scores were age- and gender-regressed, and EA2 GPS were constructed as described elsewhere[46]. We used the EA2 GPS prediction of GCSE scores as an example because the effect size of this association is currently the strongest in the behavioural sciences, accounting for 9% of the variance[46]. It will soon be possible to explain a similar amount of variance in intelligence, and with that GPS will become available to predict intelligence for individuals.

The starting point for prediction is the distribution of individual differences (see the figure, part a). The EA2 GPS is normally distributed, as GPS always are. The measure of school achievement is also normally distributed. GPS prediction of individual differences is based on its covariance with the target trait, school achievement in this example. The scatterplot between EA2 GPS and GCSE scores (see the figure, part b) indicates the difficulty of predicting individual outcomes when the correlation is modest, 0.30 in this example. Squaring this correlation indicates that EA2 GPS predicts 9% of the variance in GCSE scores. Although higher EA2 GPS can be seen to predict higher GCSE scores on average, there is great variability between individuals. For example, the individual with the second highest EA2 GPS has a GCSE score only slightly above the average. Conversely, an individual with the eighth lowest EA2 GPS has a GCSE score above the 75th percentile.

Despite this variability, powerful predictions can be made at the extremes. For example, when the sample was divided into ten equal-sized groups (deciles) on the basis of their EA2 GPS, a strong relationship between average EA2 GPS and average GCSE scores emerged that was most evident at the extremes (see the figure, part c). Specifically, the average school achievement of individuals in the lowest EA2 GPS decile is at the 28th percentile. For the highest EA2 GPS decile, the average school achievement is at the 68th percentile.

Nonetheless, individuals within the lowest and highest EA2 GPS deciles vary widely in school achievement (see the figure, part d). The overlap in the two distributions is 61%. These issues of variability in prediction are the same for any predictor that accounts for 9% of the variance in the target trait. As bigger and better GPS emerge, the predictive power will increase.

In summary, GPS are useful for individual prediction as long as the probabilistic nature of the prediction is kept in mind.

Although simple curiosity will drive consumers 'interests, GPS for intelligence are more than idle fortune-telling. Because intelligence is one of the best predictors of educational and occupational outcomes, IQ GPS will be used for prediction from early in life before intelligence or educational achievement can be assessed. In the school years, IQ GPS could be used to assess discrepancies between GPS and educational achievement, that is, GPS-based over-achievement and under-achievement. The reliability, stability and lack of bias of GPS make them ideal for prediction, which is essential for the prevention of problems before they occur. A 'precision education 'based on GPS could be used to customize education, analogous to 'precision medicine'.

A novel, socially important direction for research using IQ GPS is to understand differences within families. First-degree relatives are on average only 50% similar genetically, which means they are on average 50% different genetically. A major impact of GPS will be to recognize and respect these large genetic differences within families.

For scores on an intelligence test standardized to have a mean of 100 and a standard deviation of 15, the average difference between pairs of individuals who are selected randomly from the population is 17 IQ points. The average difference between parents and offspring and between siblings is 13 IQ points[69]. IQ GPS might help parents understand why their children differ in school achievement. Because GPS are probabilistic, a low IQ GPS does not mean that a child is destined to go no further in education than secondary school. But it does mean that the child is more likely to find academic learning more difficult and less rewarding than a sibling with a high IQ GPS.

Ethical implications

Genomic research and studies of intelligence face four principal ethical concerns: the notion of biological determinism, the potential for discrimination and stigmatization, the question of ownership of information, and the emotional impact of knowledge about one's personal genomics and intelligence. These and other ethical issues are explored in detail by the programme of ethical, legal and social implications (ELSI), which is an integral part of the Human Genome Project[70]. Also, recent books discuss ethical as well as scientific issues about personal genomics specifically in relation to education[71] and occupation[72]. Much of these ethical discussions focus on single-gene disorders, for example Huntington disease, which has 100% penetrance. By contrast, GPS are 'less dangerous 'because they are intrinsically probabilistic, not hard-wired and deterministic like single-gene disorders. It is important to recall here that although all complex traits are heritable, none is 100% heritable. A similar logic can be applied to IQ scores: although they have great predictive validity for key life outcomes[1–6], IQ is not deterministic but probabilistic. In short, an individual is always more than the sum of their genes or their IQ scores.

Issues of discrimination and stigmatization have accompanied research into genetics and intelligence from the beginning, typically because findings from both fields of study were applied to justify policies that served socio-political ideologies. For example, IQ testing was infamously used to differentiate European immigrants to the United States of America who arrived at Ellis Island in the early 1900s, and to guide eugenic ideas about sterilization in Britain and the United States of America throughout the 20th century[11]. It is important to acknowledge the risk of discrimination that occurs on the back of scientific findings about individual differences. It is, however, equally important to realize that research does not lead directly to any policy recommendations. We must be careful not to blame the scientists or entire disciplines when their findings are used wrongly[9].

Who 'owns' our genetic information? And who should decide who can access it? The question of ownership of personal data has become pivotal but also increasingly complex in our current age of information. At the same time, understanding and managing the emotional impact that stems from knowledge about our genomics and intelligence has emerged as a new societal responsibility. It is beyond the scope of this paper to elucidate these issues in the depth that they deserve but we expect that the discussions of ethical issues that surround personal genomics will consolidate the DNA revolution.

Conclusions

Genetic association studies have confirmed a century of quantitative genetic research showing that inherited DNA differences are responsible for substantial individual differences in intelligence test scores. A reachable objective shared with all complex traits in the life sciences is to close the gap between the 10% variance in intelligence scores explained by GPS and its SNP heritability of about 25%. A more daunting challenge is to break through the ceiling of 25% SNP heritability to reach the 50% heritability estimated by twin studies.

Until 2016, GPS could only predict 1% of the variance in intelligence. Progress has been rapid since then, reaching our current ability to predict 10% of the variance in intelligence from DNA alone. GPS will soon be available that can predict more than 10% of the variance in intelligence, that is, more than 20% of the 50% heritability of intelligence estimated from twin studies, and more than 40% of the 25% SNP heritability of intelligence. This is an important milestone for the new genetics of intelligence because effect sizes of this magnitude are large enough to be "perceptible to the naked eye of a reasonably sensitive observer"[73]. With these advances in the past few years, intelligence steps out of the shadows and takes the lead in genomic research.

In addition to investigating traditional issues about development, multivariate links between traits and gene-environment interplay, IQ GPS will open new avenues for research into the causes and consequences of intelligence. The new genetics of IQ GPS will bring the omnipotent variable of intelligence to all areas of the life sciences, without having to assess intelligence.

GENETIC INFLUENCE ON ATHLETIC PERFORMANCE

Lisa M. Guth and Stephen M. Roth

Abstract

Purpose of review

The purpose of this review is to summarize the existing literature on the genetics of athletic performance, with particular consideration for the relevance to young athletes.

Recent findings

Two gene variants, *ACE* I/D and *ACTN3* R577X, have been consistently associated with endurance (*ACE* I/I) and power-related (*ACTN3* R/R) performance, though neither can be considered predictive. The role of genetic variation in injury risk and outcomes is more sparsely studied, but genetic testing for injury susceptibility could be beneficial in protecting young athletes from serious injury. Little information on the association of genetic variation with athletic performance in young athletes is available; however, genetic testing is becoming more popular as a means of talent identification. Despite this increase in the use of such testing, evidence is lacking for the usefulness of genetic testing over traditional talent selection techniques in predicting athletic ability, and careful consideration should be given to the ethical issues surrounding such testing in children.

Summary

A favorable genetic profile, when combined with an optimal training environment, is important for elite athletic performance; however, few genes are consistently associated with elite athletic performance, and none are linked strongly enough to warrant their use in predicting athletic success.

Keywords: Genomics, endurance, fitness, sport

Introduction

Both the scientific and sporting communities acknowledge that genetic factors undoubtedly contribute to athletic performance. As of 2009, more than 200 genetic variants had been associated with physical performance, with more than 20 variants being associated with elite athlete status (1). Although few studies have examined the link between genetic factors and athletic performance in children or adolescents, this area of research is highly relevant to a pediatric population; the idea of predicting future athletic success through genetic testing in children is becoming increasingly common. The present review will provide an overview of the genetics of athletic performance and will focus on the relevance to young athletes.

Components of Performance

A primary challenge when attempting to describe the influence of genetic factors on athletic performance is its multifactorial nature. Every sport has unique physical requirements and these requirements can be dramatically different between sports. Therefore, any study of the genetic influence on performance must consider the performance components most appropriate for the sport of interest.

Considering the number of body systems that must interact (musculoskeletal, cardiovascular, respiratory, nervous, etc.), athletic performance is one of the most complex human traits. Perhaps the first noticeable difference between athletes of different specialties is in body morphology (i.e., height and body composition), with specific body types naturally suited to specific sports. Beyond body morphology, endurance, strength, and power are primary factors underlying athletic performance.

Aerobic endurance is the ability to sustain an aerobic effort over time, such as distance running or cycling. At the most basic level, aerobic endurance requires the ability of the cardiovascular system to deliver oxygen to the working muscles and the ability of the muscles to utilize that oxygen. The most common quantification of endurance is the maximal rate of oxygen uptake (VO_{2max}). However, VO_{2max} does not perfectly correlate with endurance performance (e.g. marathon running), as other factors such as economy and ventilatory threshold also influence performance.

Muscular strength is the ability of the muscle to generate force. Muscular strength is generally quantified by the one repetition maximum. Muscle power is the interaction between the force and velocity of a muscle contraction (e.g. an explosive movement such as vertical jump). Muscle strength and power are critical in athletic events such as sprinting, jumping, and weightlifting.

Additional components of athletic performance include cognitive factors and injury susceptibility. It is critical to remember that the environment (e.g., training, nutrition) also influences many of these traits. An individual's "trainability," or response to exercise training, is also partially dependent on genetic factors, as recently reviewed by Bouchard (2). The relative importance of environment versus genetic factors on athletic success likely varies widely between sports as well (i.e., gymnastics vs. 100m sprint). Elite athletic status, therefore, results from the interaction of an optimal combination of genetically driven physical and mental traits with the ideal environment for athletic success (3).

Heritability of sub-traits

The heritability of a trait is generally considered an estimation of the importance of genetic factors to that trait. For example, the heritability of athletic status (regardless of sport) is estimated to be 66% (4). Height, which is critical for success in some sports, is highly heritable, with about 80% of the variation due to genetic factors (5). Body type (having mesomorphic or ectomorphic somatotype) is also highly heritable (6). These somatotypes are classically associated with power or endurance athlete status, respectively (7).

Costa et al. (8) recently reviewed the existing family and twin studies related to specific endurance and muscular strength phenotypes. Aerobic endurance, as reflected by VO_{2max} has a heritability of about 50% (9). Heritability estimates for muscular strength, and power range from 30 to 83%, depending on the specific muscle and type of contraction (8).

Key Performance Genes

Though many specific genes and sequence variants (polymorphisms) within genes have been associated with performance, many of the findings to date have not been adequately replicated. Two notable exceptions are the angiotensin-1 converting enzyme insertion/deletion (*ACE* I/D) polymorphism, and the α-actinin-3 (*ACTN3*) R577X polymorphism, both of which have been examined in several populations using a variety of experimental approaches.

ACE I/D

Fifteen years ago, the *ACE* I/D polymorphism was the first genetic factor to be associated with human performance (10). The *ACE* gene codes for angiotensin-1 converting enzyme, part of the renin-angiotensin system responsible for controlling blood pressure by regulating body fluid levels. The *ACE* I allele represents a 287 bp insertion and is associated with lower serum (11) and tissue (12) ACE activity while the D (deleted) allele is associated with higher serum and tissue ACE activity (13). The *ACE* I/I genotype is consistently associated with endurance performance and higher exercise efficiency while the D/D genotype is associated with strength and power performance, though some conflicting reports do exist (13). Notably, there is no association between I/D genotype and elite athlete status in Kenyans (14), highlighting the potential confounding factors of ethnicity and/or geography. An extensive review of the existing literature on *ACE* genotype and athletic performance through 2011 is available elsewhere (13). Additionally, a systematic review and meta-analysis of 25 studies examining the association of *ACE* genotype with sport performance was recently performed by Ma et al. (15). Overall, the *ACE* I/I genotype was specifically associated with performance in endurance, but not power, athletes, supporting the general consistency in the literature for an association of *ACE* I/D genotype with endurance performance.

ACTN3 R577X

The *ACTN3* gene codes for the protein α-actinin-3, a structural sarcomeric protein found exclusively in the fast type II muscle fibers used during explosive activities. A polymorphism leads to a premature stop codon (X) rather than an arginine (R) at position 577. The R allele is generally considered to be advantageous in power-oriented events, as the RR genotype is overrepresented in elite power athletes (16) while the XX genotype is associated with lower sprinting ability and muscle strength (17). The *ACTN3* R577X variant was recently studied across three groups of elite European athletes (633 athletes and 808 controls). In line with previous literature, power athletes were approximately 50% less likely to have the XX genotype and endurance athletes were approximately 1.88 times more likely to have the XX genotype vs. the RR genotype. Interestingly, for endurance athletes, the odds ratio for having the XX genotype was about 3.7 times larger for world-class compared to lower competition level athletes, suggesting the ACTN3 genotype may be even more important at the highest levels of performance (18).

A recently published meta-analysis of 23 studies examining the association of *ACTN3* with sport performance (15) demonstrated increased probability of performance in power events in R carriers, supporting the general consistency in the literature of the association between the *ACTN3* genotype and power-oriented athletic performance. The association of the *ACTN3* R577X variation with performance is arguably the strongest such association to date. Genotype frequencies have not only been repeatedly linked to athlete status and performance phenotypes, but experimental animal models also support the detrimental effect of α-actinin-3 deficiency on muscle performance. It is important to note that the most consistent associations between the *ACTN3* genotype and performance have been observed in athletes; these associations have been recently reviewed by Eynon et al. (19). A 2011 meta-analysis also supported higher prevalence of the RR genotype in sprint and power athletes but found no association of *ACTN3* with physical capabilities in the general population (20), thus the importance of *ACTN3* on muscle function in the general population remains somewhat unclear.

Injury Risk

Resistance to and/or the ability to recover from injury is another critical factor for optimal performance. Two main injury types have been studied with regard to genetic risk: concussion and tendinopathies. These areas of research are particularly important to the skill development of the pediatric athlete, as injuries can dramatically decrease time spent in training. Additionally, some injuries can result in recurrent issues throughout life. For example, athletes who sustained concussions decades earlier exhibited more brain anomalies and greater cognitive decline with aging compared to athletes that had never been concussed (21). A better understanding of the genetic components to injury risk and recovery could improve our ability to protect at-risk young athletes from serious injury and to optimally treat the injuries that do occur.

Concussion

The gene most frequently studied with regard to concussion/mild traumatic brain injury has been *APOE*. *APOE* has three isoforms (ε2, ε3, and ε4 alleles) and its ε4 allele has been associated strongly with Alzheimer's disease over the past decade (22). Based on this association, several groups have begun to evaluate the association of the ε4 allele with risk for concussion and outcomes from traumatic brain injury, though the research to date is unclear. For example, individuals with the ε4 allele suffered worse outcomes from head injury (23) and boxers possessing an ε4 allele had higher chronic brain injury scores (24), consistent with the idea of ε4 as a risk allele. However, a prospective study of collegiate athletes did not find an increased risk of concussion in athletes carrying the ε4 allele (25) and the *APOE*-ε4 genotype was not associated with poorer outcomes following mild traumatic brain injury in children (26). A separate *APOE* variant (G-219T) has been retrospectively associated with concussion risk in athletes, with 3-fold higher risk in athletes with the TT genotype compared to the GG genotype (27). The same study also identified a weak association between the τSer53Pro polymorphism in *MAPT*, the tau-protein encoding gene, and concussion risk.

Tendinopathy

With regard to tendinopathies, collagen is the primary structural component of tendons and ligaments. Unsurprisingly, variants in two collagen-encoding genes (*COL1A1* and *COL5A1*), a gene involved in connective tissue wound repair (*MMP3*), and the gene encoding *TNC*, an extracellular matrix protein, have all been linked to increased risk for tendinopathy (28,29). The presence of multiple risk alleles appears to further increase injury risk (30). As with most areas of genetics and performance, these studies are among the first to provide evidence for association and require considerable replication and validation.

Relevance for Young Athletes

Few studies have examined the association between genetic variation and athletic performance phenotypes in child or adolescent athletes. This is unsurprising given the potential ethical considerations of genetic testing in children. The *ACE* II polymorphism was reported to be associated with significantly greater handgrip strength and vertical jump performance in female, but not male Greek adolescents (31), which contrasts the typical association of the D allele with better strength and power in adults. Recently, the *ACE* D allele has been associated with greater handgrip strength in adolescent girls (32) and standing long jump performance in middle-school age children (33).

The association of the *ACTN3* genotype with performance has also been studied in children. Boys with *ACTN3* RR genotype tended to swim faster (25m and 100m) (34) regardless of training status. The R allele was also associated with better 40m sprint performance in adolescent Greek boys (35). Late adolescent girls with the RR genotype performed better on sit up tests than girls with the RX genotype (32). However, the *ACTN3* genotype was not associated with several other power or endurance phenotypes in adolescents of either sex (35). Additional correlations between *PPARA*, *PPARD*, and *PPARGC1A* genotypes with standard fitness tests in children were also reported (32,33). Overall, many of these studies associating genetic variation with performance in children have been underpowered and have failed to correct for multiple comparisons, so it is premature to draw firm conclusions. Further, the drive behind the prediction of athletic performance with genetics is primarily aimed at the early identification of individuals who will become exceptional athletes as adults.

Genetic Testing

Talent selection, the identification of promising athletes at a young age allows for an earlier adoption of specialized/ dedicated training. Historically, talent identification has been based on physical and psychological characteristics and sport-specific performance. Genetic testing may provide an additional way to predict adult performance traits prior to their full development in untrained children by profiling combinations of gene variants associated with a particular trait. As DNA sequence is constant throughout life, genetic testing can be performed as soon as a DNA sample is available (in infancy or even prior to birth).

Despite the lack of performance predictability offered by single genetic variants, several companies are marketing genetic tests claiming to do just that. Tests are available for frequently studied genes (e.g., *ACTN3* and *ACE*) as well as several genes with relatively little scientific evidence (36). In fact, a genetic test for *ACTN3* was developed in 2004, only one year after Yang et al. (16) first described its potential connection to sport performance. Now this test is being marketed directly to coaches and parents with no prescription required (37). Based on this single test, this company will interpret an individual's "genetic advantage" as "predisposed to endurance events," "predisposed to sprint/power events," or "equally suited for both endurance and sprint/power events." This is by no means the only available direct-to-consumer genetic test, but is a representative example of the existing commercial options.

Beyond the oft-insufficient rationale for testing these variants, most coaches, parents, and athletes lack the scientific background required to understand the limitations of these tests or the implications of the results. However, some professional sports teams are already using the results of these tests to partially direct training prescriptions (38,39).

It remains to be seen whether the complex contribution of genetic factors to athletic performance can be used to improve talent selection. It is important to remember that genetic association studies reveal factors that are associated with athletic performance traits at a population level and the relative importance of any given variant for an individual is undoubtedly more variable, thus genetic screening cannot be used to conclusively predict or rule out athletic success (40). There is neither currently nor is there likely to be a gene variant that is either required or sufficient for superior athletic performance.

The potential for genetic testing to predict injury susceptibility, such as *APOE* genotype with response to concussion, may provide a unique and important avenue to improve safety for athletes. Though consensus remains elusive, the existing evidence points to the idea that genetic factors will be identifiable as important for injury susceptibility and could potentially be used to the advantage of young athletes as they consider sport participation. But, the ethical challenges related to genetic testing in relation to sport performance, especially in children, are especially difficult, as reviewed recently by Wackerhage et al. (41), and need to be considered carefully.

Conclusion

Current evidence suggests that a favorable genetic profile, when combined with the appropriate training, is advantageous, if not critical for the achievement of elite athletic status. However, though a few genes have now been repeatedly associated with elite athletic performance, these associations are not strong enough to be predictive and the use of genetic testing of these variants in talent selection is premature.

TECHNOLOGY EQUIPMENT RECAP

HydroGenes - 1! Proton and 1! Electron.

Melanocytes - Photovoltaic Cells

Neurons/Nerves - Electromagnetic Cells

Fascia - Plasma Medium (misonomered Ether)

Brain - CPU, Inductor

Brainstem - Two-way Adapter for CPU into the Motherboard

Pineal Gland - Receiver, Crystal Tuner & Actuator Arm/ Head responsible for Phosphorescence, Thermoluminescence, Piezoelectricity, Birefringence & Harmonic Generation (very much like the otoconia in the ears)

Operating System - Deductive Logic or PQ

Heart - Hydraulic Ram, Turbine (from the Greek τύρβη, tyrbē, or Latin turbo, meaning vortex) and Hard Drive.

Melanosomes - Alternators

Mitochondria - Motors

Myelin Sheath - Insulation

Cytoskeleton - Filaments

Phospholipids - Capacitors, Dielectric Material (lipids in general)

Spine - Piezoelectric, Motherboard, Radio Wave Antenna

RBC - Floppy Discs

Lymph Nodes - Filters, Nodes

Tastebuds - Electronic Scanners

Protein - Transformer

Transformers 'Roll Out' - Conformational Change (Macromolecule Shape Shifting)

Antioxidants - Semi-conductors (especially the selenium based...)

Body Cells - Plasma based Crystal disc, fitted with integrated circuits as well as gates and channels (see Human Cell Membrane and/or Computer Chip)

Nerves & Vessels - 'Copper' wires (CoAxial Cables) and Fiber Optics

Pigment, Nerve & Blood Clusters - Input Devices like a Mouse, Keyboard, etc...

DNA - Piezoelectric, Antenna, Data Storing Inductors.

DNA sub entry **Tissues** - Short Living Stories.

DNA sub entry **Genome** - Substrate & Product, a digital Library (HardDrive) of all your Ancestors have ever seen, said, touched, tasted or heard.

DNA sub entry **Chromosome** - Rewritable Unlimited Storage Books (Folders) of the Library, DNA.

DNA sub entry **Histone** - Writing instrument, encoders and **book spines.**

DNA sub entry **non-coding RNA** - Self Organizing Books Shelves

DNA sub entry **Gene** - Chapters (Files) in the Books, source codes.

DNA sub entry **Messenger RNA** - Protein Information, a Sentence.

DNA sub entry **Codon** - Word (Binary Code there are 2 bonds between each 3 nucleotides representing their arrangement), Amino Acid.

DNA sub entry **Nucleotide** - Letter

DNA sub entry **Nucleoside** - Bits of Information

Collagen Based Tissue - Piezoelectric Inductors

Stomach - Chemical Mixer

Lumen - the SI unit of luminous flux = to the amount of light emitted per second..... or hollow structures in vessels and cells... hmmm????

Eyes - Camera Lens/Charge Coupled Device (CCD), Digital to Analogue Converter, Complex Photovoltaic Cells/ Photodetector...

Amino Acids - Fuses that can be almost anything!

Nucleic Acid - Actual Intelligence (self powering too).

N-Type Semiconductors - Selenium or Silica doped with Phosphorus (Alkaline-ish)

P-Type Semiconductors - Selenium or Silica doped with Boron (Acid-ish)

PN Junction - Crystal Lattice Structure Material allowing the flowing of electrons in one direction.

Bone and Fascia seem to be a massive N-Type, P-Type, PN Junction Super computer on it's own... especially if we add in the Piezoelectricity & Vitamin D!

Rectifier - N-Type + P-Type + PN Junction in Bone

Melanin - CPU Core, Solar Repeater

Human Cell Membrane and/or Computer Chip - A flat semiconducting (crystal) disc or wafer, with integrated circuits (resistors/conductors) and/or gates & channels. We now have to add the filaments into this Crystal Disc we call a Body Cell or Somatic Cell.

Transistors - a semiconductor device with three connections, capable of amplification in addition to rectification.

The location that a virus goes viral in, is called a **Hotspot**? WTH!

Virus - an infective agent that typically consists of a nucleic acid molecule in a protein coat, is too small to be seen by light microscopy, and is able to multiply only within the living cells of a host.

Wait you see that, it is happening again! Host...

See look there is another definition of **Virus** - a piece of code that is capable of copying itself and typically has a detrimental effect, such as corrupting the system or destroying data.

Wait a damn minute! DNA is a piece of **code**... A viral strand of DNA or RNA that can jump host is fully capable in that context of copying itself, one would even argue, that is it's only 'motion'. The detrimental effects of corrupting the system (physical illness) or destroying data (mental illness), can clearly be seen anthropomorphically.

Host - an animal or plant on or in which a parasite or commensal organism lives. VS

Host - store (a website or other data) on a server or other computer so that it can be accessed over the internet.

Transmission is the act of transferring something from one spot to another, like a radio or TV broadcast, or a disease going from one person to another.

I am highlighting the unknown and proposing we may have some answers! Infection - an infectious disease.

plural noun: infections "a chest infection"

Vs

Infection - the presence of a virus in, or its introduction into, a computer system. What is a computer system?

Computer System - a computer system is a programmable electronic device that can accept input; store data; and retrieve, process and output information.

Pandemic language = Virology/Biology language. The question is, why? The next question is what does that have to do with Dr. Sebi or Robert Becker? The obvious....

Computer System - a computer system is a programmable electronic device that can accept input; store data; and retrieve, process and output information.

Computer System - a single information processor but usually a group of processors that have specified and general computations; grouped by hardware ie... liver cells, lung cells, brain cells etc.. What you think?

Exercise - activity requiring physical effort, carried out to sustain or improve health and fitness.
"exercise improves your heart and lung power"

Exercise - computer training or computer based training.

Exigenetics - Term created by Dr. EnQi for Melanin vs Diabetes research, denoting the control that exercise has over gene expression.

Hydration - the process of inducing gelling, ionizing, dissolution & turbulent flow with activation of cytochrome c (via infrared light).

Resonance - the quality in a sound of being deep, full, and reverberating. "the resonance of his voice"

- The ability to evoke or suggest images, memories, and emotions."the concepts lose their emotional resonance"

- The reinforcement or prolongation of sound by reflection from a surface or by the synchronous vibration of a neighboring object.

- The condition in which an electric circuit or device produces the largest possible response to an applied oscillating signal, especially when its inductive and its capacitative reactances are balanced.

- The condition in which an object or system is subjected to an oscillating force having a frequency close to its own natural frequency.

- The occurrence of a simple ratio between the periods of revolution of two bodies about a single primary.

- The state attributed to certain molecules of having a structure that cannot adequately be represented by a single structural formula but is
a composite of two or more structures of higher energy.

- • A short-lived subatomic particle that is an excited state of a more stable particle.

Induction - the action or process of inducting someone to a position or organization."the league's induction into the Baseball Hall of Fame"

Induction - a formal introduction to a new job or position.plural noun: inductions
"an induction course"
enlistment into military service.

Induction - The process or action of bringing about or giving rise to something."isolation, starvation, and other forms of stress induction" the process of bringing on childbirth or abortion by artificial means, typically by the use of drugs.

Induction - The inference of a general law from particular instances.

Induction -"the admission that laws of nature cannot be established by induction" the production of facts to prove a general statement.

Induction - a means of proving a theorem by showing that if it is true of any particular case it is true of the next case in a series, and then showing that it is indeed true in one particular case.

Induction - noun: mathematical induction; plural noun: mathematicals inductionthe production of an electric or magnetic state by the proximity (without contact) of an electrified or magnetized body.

Induction - The production of an electric current in a conductor by varying the magnetic field applied to the conductor.

Induction - The stage of the working cycle of an internal combustion engine in which the fuel mixture is drawn into the cylinders.

Is there anyone reading this that would disagree with our body fitting these definitions, the definitions of a computer?

Man this thought experiment just got a lot more interesting didn't it? MIT and the US Military are different types of receipts huh? Is it possible frequency resonance, spreads disease? Human modems? Can Shedding be a broadcast signal?

Wi-Fi is a wireless networking technology that uses radio waves to provide wireless high-speed Internet access. A common misconception is that the term **Wi-Fi** is short for "wireless fidelity," however Wi-Fi is a trademarked phrase that refers to IEEE 802.11x standards.

Viral shedding is a term for when viruses are replicating or reproducing, the virus is being led out of the host cell where it's replicating or
reproducing ... Viral shedding is the expulsion and release of virus progeny following successful reproduction during a host cell infection. Once replication has been completed and the host cell is exhausted of all resources in making viral progeny, the viruses may begin to leave the cell by several methods.

Vaccine - a substance used to stimulate immunity to a particular infectious disease or pathogen, typically prepared from
an inactivated or weakened form of the causative agent or from its constituents or products.

Vaccine - a program designed to detect computer viruses and inactivate them.

"the rate of use of vaccines for computer viruses is not as high as in the US, Japan, and other countries"

Application - a medicinal substance put on the skin.

Application - a program or piece of software designed and written to fulfill a particular purpose of the user.

In our thought experiment, if a virus is simply the media for harmful information...

Media - an intermediate layer in the wall of a blood vessel or lymphatic vessel.

Media - the main means of mass communication (broadcasting, publishing, and the internet) regarded collectively.

DOPE - an illicit drug (such as heroin or cocaine) used for its intoxicating or euphoric effects
especially : MARIJUANA (dopamine altering)

Dope - a preparation (such as an anabolic steroid, diuretic, or tranquilizer) given to a racehorse to help or hinder its performance

To Dope - In semiconductor production, to dope is the intentional introduction of impurities into an intrinsic semiconductor for the purpose of modulating its electrical, optical and structural properties. The doped material is referred to as an extrinsic semiconductor.

Short Circuit - Cardiac Arrest?

Short Circuit - Multiple Sclerosis (due to loss of insulation)

Overheating - Fever?

Overcurrent - Inflammation

With Infection and Virus included we are onto something.

Current - belonging to the present time; happening or being used or done now.

Current - *a body of water* or air *moving in a definite direction*, especially *through a surrounding body of water* or air in which there is less movement.

Current - a flow of electricity that results from the

ordered directional movement of electrically charged particles.

Current - a quantity representing the rate of flow of electric charge, usually measured in amperes.

Current - the general tendency or course of events or opinion.

Leakage Current - the unintended loss of energy, gain of resistance or results of faulty/worn out insulation.

Plasma - Electric Currents or Electric Current Carrier

Electric Current - Magnetic Field (AtomSphere) Carrier

Alternating Magnetic & Electric Waves - Light

NeuroTransmitters - Record of ElectroMagnetic Waves produced by Neurons (ElectroChemical Message)

Hormones - Large simple versions of NeuroTransmitters (ElectroChemical Message)

Malware - External Negative Mental Programming

Food - Informative Electronic Batteries

0) Movement and sound create energy from water for basic cellular function, via the EnQi Cycle which includes Mitochondria Water. This system slowly increases as all other energy systems fail.

1) Phosphocreatine - anaerobic (no respiration required), phosphocreatine donates it "phospho" to ADP to recycle ATP. This makes 10 ATP per second, its a 1 to 1 ratio (1 phosphocreatine creates 1 ATP) and this is the jump start energy.

2) Anaerobic Glycolysis - anaerobic (no respiration required), Glycogen &/or Glucose to Lactate, 5 ATP per second, 1 to 3 ratio (1 Glycogen creates 3 ATP while 1 Glucose creates 2 ATP) and this is bulk of the energy we focus on, 9 - 120 seconds.

3) NAD/Cytochrome 1 - aerobic (requires oxygen), Glycogen &/ or Glucose to CO_2/H_2O, 2.5 ATP per second, 1 to 38 ratio (1 Glycogen &/or Glucose creates 38 ATP), 2 minutes up to 2 hours.

4) FAD/Cytochrome 2 - aerobic (requires oxygen), FFA &/or Triglycerides to CO_2/H_2O, 1.5 ATP per second, 1 to 360 ratio (1 Glycogen &/or Glucose creates 360 ATP), 2 minutes up to 2 days.

Food rule of thumb - Resynthesis of ATP of Inverse to Yield, the closer the ratio is to 1:1 the fast it can be recycled.

Muscle rule of thumb - Frequently used muscle is slow twitch, Fast twitch is slowly used (at that's the blueprint).

Electric Power - the **rate** at which work is done or energy is transformed into an electrical circuit. Simply put, it is a measure of how much energy is used in a span of time.

Conductor - a person who directs the performance of an orchestra or choir.

Conductor - a material or device that conducts or transmits heat, electricity, or sound, especially when regarded in terms of its capacity to do this.

Lymphatic System - Watermill

Circulatory System - Generator

Integumentary System - Photovoltaic Diaphragm

Immune System - Antivirus, Malware Scanner, Frequency Filter & Rectifier

Nervous System - Power Transmission and Cellular Communications Lines

Fascia System - HydroElectric Grid, Scaffolding

Respiratory System - Windmill

Windmill - a structure that converts wind power or "air" power into rotational energy or vortex energy, to mill grain. In our case grain is Magnetism!

MAGNETS ARE DEFINED BY GRAINS
MAGNETIC GRAINS ARE DEFINED BY APPLIED STRESS AND CRYSTAL GEOMETRY
SPM SUPERMAGNETIC
SD SINGLE DOMAIN
PSD PSEUDO DOMAIN
MD MULTIDOMAIN

Reproductive System - Quine (self-replicating programs)

Skeletal System - Piezoelectric Crystal Shaped to produced highly specific frequency under stress, Dynamic Oscillators.

Urinary System - Industrial Wastewater, Return Flow, Surface Runoff, Urban Runoff Agricultural & Animal Husbandry Wastewater

Digestive System - Massive Inductor

Mouth - Industrial Grinder

Endocrine System - Programmer for Human Cell Membrane and/or Crystal Gel Computer Chips

Human Being - Resonator

Vessels - Pipes

Aromatic Ring - Cyclotron (Particle Accelerator)

Glycation - Corrosion

Exegenetics - Holistic Biomechanics; the purposeful science of combining light, water, diet & exercise to effect DNA.

EnQi's 1st Law of Metabolism - The conversion rate of cholesterol should match the activity of Melanin in the skin.

These two systems are designed to be and stay coupled. A dark skin person with low sunlight intake and low exercise is going to die from a Metabolic Complication. The only time Animal Flesh is safe to be consumed by a Eumelanin Dominant person is in times of starvation or extremely high activity.

This Law is a Constant and when broken results in disease every time.

EnQi's 2nD Law of Metabolism - The average rate of applied mechanical stress on the bone electrically stimulating bone marrow, determines the rate of bone deterioration and Red Blood Cell production.

EnQi's 3rd Law of Metabolism - The human body metabolizes Transverse Waves and Mechanical Waves into Electricity. Electricity is the main driver of Biochemistry. Exercise is just as potent a driver of Biochemistry as the Sun.

EnQi's 4th Law of Metabolism - Electron movement and bonding is the Nature of Chemistry. PhotoChemistry and PiezoElectroChemistry are the Primary drivers of Biochemistry.

The Ancients discovered this and created Martial Artforms as a way to clean the Bone, Bone Marrow & Brain. Plaque & Sugar are the top drivers of Brain Disease. The things destroying the Heart are secondarily destroying the brain, and they are the breaking of these Universal Laws.

EnQi's 5th Law of Metabolism - Nutrients are actually substrates that must be transformed via biochemistry to be meaningful. This means that providing your body with lots of nutrition without the Water, Light & Exercise don't work alone.

EnQi's 6th Law of Metabolism - The Body maintains the least amount of bone marrow required to handle blood demand. The marrow is very energy demanding, thus attracting and storing fat for energy, eventually becoming fat itself. Fatty bone marrow is called yellow bone marrow. Yellow Bone Marrow can be reconverted to Red Bone Marrow should the body's demands require it, and the body's resources facilitate it. The primary driver is pressure, hormesis training on the Bones. BMR is heavily driven by Bone Marrow, this means Bone Marrow is a driver if insulin and insulin resistance.

EnQi's 7th Law of Metabolism - The system of pigments throughout the body are for metabolism of Light, actual Soulfood. The Adsorption & Absorption of Photons by Water.

Adsorption - increase in the concentration of a dissolved substance at the interface of a condensed and a liquid phase due to the operation of surface forces.

Absorption - a physical or chemical phenomenon or a process in which atoms, molecules or ions enter some bulk phase – liquid or solid material. This is a different process from adsorption, since molecules undergoing absorption are taken up by the volume, not by the surface.

EnQi's 8th Law of Metabolism - in a diabetic state, sugar is simply invisible to the body. Sugar is not being "sensed" because it's not being converted to energy. In this state of starvation the body turns on every pathway it has to produce sugar from everything you have in your body, fats and proteins included.

This is the reason that it seems like no matter what you eat or 'don't eat', your blood sugar goes up. It's very frustrating. The only way to make it stop is converting that substrate (glucose) into it's final product (energy). The reception of the actual energy, tells the body to stop producing substrate, we good. This must start in the legs and back, the largest muscles but most overlooked. The legs are particularly punished by sitting for extended periods of time, 3-6 hours straight, for a total over 3/4 the time your awake! The leg circulation atrophies and destroys the nerves, nerves are neurons that need a lot of nutrients!

*You must cross reference any and all protocols; food, exercise, medication etc... with the Constitution book & Declaration of Independence!

www.ingramcontent.com/pod-product-compliance
Lightning Source LLC
Chambersburg PA
CBHW071913210526
45479CB00002B/403
* 9 7 9 8 3 3 4 6 1 7 2 3 0 *